W9-AOV-291

UA
UA990 .P4

Terrain Evaluation

Robert Manning Strozier Library

SEP 9 1974

Tallahassee, Florida

Terrain Evaluation

An introductory handbook to the history, principles, and methods of practical terrain assessment

Colin W. Mitchell
(Lecturer in Geography at the University of Reading)

with a Foreword by

J. M. Houston
Principal of Regent College, Vancouver

Longman

LONGMAN GROUP LIMITED
London
*Associated companies, branches and representatives
throughout the world*

© Longman Group Limited 1973

All rights reserved. No part of this publication may be reproduced, stored in a retrieval system, or transmitted in any form or by any means, electronic, mechanical, photocopying, recording, or otherwise, without the prior permission of the Copyright owner.

First published 1973

ISBN 0 582 48426 X Limp
   0 582 48425 1 Cased
Printed in Great Britain by
Lowe & Brydone (Printers) Ltd., Thetford, Norfolk

To Alec Barrie

Contents

List of illustrations

Foreword

by Dr J. M. Houston, Principal of Regent College, Vancouver

It is typical of most disciplines that they consist of both a formal and a functional aspect. In geography, we associate these two understandings in terms of regional and systematic studies respectively. Like the wings of a bird, a discipline will lose its flight if it does not balance these understandings. Both need to be exercised. Regional studies of landscapes need to be complemented by the new techniques being developed rapidly, to appraise, measure and evaluate the changes wrought by man. There is therefore a real need for handbooks that cover succinctly the technical issues involved in the professional understanding of landscapes. This book is an outline of the terrestrial geometry of the globe, outlining the techniques required to understand and appraise its features.

It is a new science that embraces much more than the original concern of the terrestrial sciences such as geography, geology, climatology, plant geography, and others. Its application today to the soil scientist, agriculturist, regional planner, conservationist, fiscal controller, military strategist, and other professionals is becoming apparent. Furthermore, because of the pragmatic issues of environmental deterioration, world hunger and the population explosion, and other related issues, we can no longer afford to persist in our ignorance concerning vast stretches of Africa, Asia and Latin America. Our spaceship Earth now needs to be known much more intimately, simply in order to survive into the future. The Biological International Programme is committed to mapping the world's ecosystems. The evaluation, classification, and mapping of the earth's terrain is a necessary first step in this direction.

We are leaving behind us an age that has viewed nature as an objective reality. Today, we are living in a more pragmatic, less idealistic age. The nouns of previous thought forms are the verbs of a new outlook. Pure thought is being replaced by the use of knowledge for a technological society, busily engaged in changing the face of the earth. It is an age of operational thinking. The very functions of nature are being threatened by scientific takeovers, and it is not too fanciful to predict that the daily weather, the biological cycles of life processes, as well as the energy of the atom will become harnessed to human corporations. Thus it becomes imperative that all thoughtful citizens should know something of the changes man has already wrought in his physical habitat, and which he is now modifying with accelerating power.

There is another issue that affects also the future of our disciplines, and specifically regional sciences, including geography. It is the debate between *realism* and *nominalism*, first raised by Aristotle: the debate whether the 'ideal' has a real existence, of which individual examples are its expressions; whether the nominalist view is the correct one, that abstractions are only abstractions. Dr Mitchell takes his stance as a nominalist, and his book is therefore orientated towards the identification of each unit of the earth as a unique unit, though mappable and classifiable. With the advent of computer science, capable of mapping the earth's terrain, the techniques elaborated in this book are a first step towards this realisation. It is a book to be commended to students of geography, conservation, ecology, land use, engineering, military science, and many other specialists who are engaged in understanding the face of the earth for particular objectives.

J. M. HOUSTON

Preface

The aim of this book is to meet the need, expressed by T. G. Miller (1967) and others, for a brief, popular and inexpensive introduction to terrain evaluation. The work done by many individuals and organisations over the past fifteen years has caused the field to expand almost beyond recognition. Although some workers have published summaries of their own and related contributions (e.g. Christian and Stewart, 1964; Beckett and Webster, 1969, 1970), and an important symposium assembled summaries of the most recent advances (Stewart, 1968), there is still a need for a simple collation and summary for the benefit of those most concerned: earth scientists, those whose professions involve the use and management of land, university students, and general readers.

There are other reasons for attempting this task now. In a few years' time, the developments promise to be so extensive that a summary of the present scope may neither be possible for a single individual, nor feasible in a short book. Moreover, terrain evaluation is still a subject composed of a number of concepts and techniques which have not attained the organic unity of a single academic discipline. This book aims to help the field towards this unity and towards attaining a distinct place within the ambit of geography.

The author is conscious of the fact that there is no aspect of the subject, save the single possible exception of the interpretation of arid terrain on aerial photographs, at which there are not others considerably better qualified and expert than himself. He would nevertheless maintain that there is distinct value in the book being written by one person, to give unity not only of plan and conception, but also of treatment. An attempt has been made to systematise the whole range of methods and disciplines required even at the expense of treating them in a manner which may appear elementary and superficial to specialists.

Acknowledgements

First of all, I should like to thank my friend and former tutor Dr J. M. Houston for asking me to write this book and encouraging me to do so; without this it would probably never have been undertaken.

Most of the information on which it is based was obtained while I was working, first at Cambridge and subsequently at Reading Universities, under the direction of Dr R. M. S. Perrin, on a research project for the Military Engineering Experimental Establishment, and I am indebted to many members of the staffs of all three institutions for the backing and help they gave.

I gratefully acknowledge the assistance of the following: Professor P. Hall and Professor R. A. G. Savigear for facilities and encouragement, Dr P. H. T. Beckett, Dr J. R. Hardy, Mr M. Parry, Dr R. M. S. Perrin, Dr R. Webster, Dr J. B. Whittow, and Dr S. G. Willimott for comments on the outline and the manuscript, and among these I am especially indebted to Dr Perrin for overseeing the section on laboratory analyses in chapter 17 and to Dr Webster for writing chapter 18. I also thank Mrs J. Gillo and Mrs J. Wild for the typing, and my wife for her unfailing patience and encouragement.

I am indebted to the Ministry of Defence (Air Force Department) for Figs. 15.1 and 15.2.

Part One

Principles of Terrain Evaluation

1
Introduction

Definition of terms

The term *terrain evaluation* has been adopted following the precedent of previous research carried out under the auspices of the Military Engineering Experimental Establishment (Beckett and Webster, 1969).

It has developed in response to the need for an understanding of terrain by an increasing variety of disciplines concerned with its practical uses. These are both scientific, such as geology, hydrology, geography, botany, zoology, ecology, pedology and meteorology; and applied, such as agriculture, forestry, civil and military engineering, and urban and recreational landscape design. This range of interest makes it important to be especially careful about the terms employed.

Terrain is defined by the *New English Dictionary* as a 'tract of country considered with regard to its natural features and configuration'. This is preferable to other similar terms because its meaning is more strictly confined to the surface of the earth, has fewer academic and practical connotations, and has already been used in publications in this sense (Beckett and Webster, 1969). *Environment* and *milieu* have meanings which are somewhat too general and extend well beyond the confines of geography; *physiography* is an older term which includes not only surface form and geology but also climatology, meteorology and oceanography, and indeed natural phenomena in general. *Geomorphology* has the advantage of being more narrowly confined to landforms but is too strongly involved with considerations of process. *Microrelief* is too exclusively geometric and does not comprehend earth materials or structure. *Soil*, conversely, is too largely concerned with material and too little with geometry. *Landscape* or *land* are perhaps the closest equivalents, but both are somewhat wider concepts than terrain, and the former rather too strongly connotes the visual and artistic aspects. The evaluation of landscape is an important and growing aspect of earth science, but only occupies a small place in this book because aesthetic factors have a more tenuous and less easily defined relation to the detail of the ground surface. This has retarded their quantification and appraisal in comparison with other aspects of terrain study.

Evaluation is defined as the 'act or result of expressing the numerical value of; judging concerning the worth of' an object. This double meaning makes it somewhat more inclusive and thus preferable to such terms as *analysis, classification, quantification, assessment,* or *appraisal.*

Terrain evaluation, as a process, involves first *analysis*: the simplification of the complex phenomenon which is the natural geographic environment; secondly *classification*: the organization of data distinguishing one area from another and characterizing each; and thirdly, *appraisal*: the manipulation, interpretation, and assessment of data for practical ends.

The need for terrain evaluation

The need for a 'bank' of readily available data about terrain over the world has been felt for a long time. Whenever agricultural, engineering, military, or other works are to be carried out in an unknown area it is necessary to start from the basis of available published information in the form of books, journal articles, maps, and aerial photographs. Much more data usually exists in unpublished files and notebooks not readily available or adequately indexed for the new enquirer and stored according to what in the Army would be called the 'heap system'. Still more data are unwritten and either lost altogether or existing only in the memories of living individuals whose existence or whereabouts is unknown or at least undocumented. There is therefore a special need for an information store about terrain which would not only enable an enquirer to benefit from past experience but also assist him in the acquisition of new information. All types of land user would be interested in such a store, which would abstract, sort, and classify practically important data about terrain in such a way that it was readily and rapidly available.

The scope of terrain evaluation

The scope of the subject is wide. It begins with the user's need and the whole problem of acquiring and classifying old information about the terrain and its practical uses from the sources previously mentioned and acquiring new knowledge from field surveys and associated laboratory work and statistical analyses. It therefore includes the study of these techniques. Secondly, it is concerned with the abstraction, classification, and storage of such information to make it available quickly, cheaply and efficiently. Thirdly, it considers the means by which such information is retrieved, reproduced, and supplied to users in an accurate and comprehensible manner.

Terrain evaluation, therefore, has a basis in all the pure sciences concerned with the surface of the earth's crust, and all the applied sciences concerned with its uses. In addition, it involves both the theory and practise of data acquisition, processing and communication.

There are, however, three main types of phenomenon which are generally excluded:

1. The atmosphere. This is too variable and ephemeral to be assigned to sufficiently small and closely definable tracts of the earth's surface.

2. Permanent expanses of water. These are not terrain in the strict sense.

3. That part of the earth's crust which lies at a depth greater than about six metres. Terrain evaluation cannot be concerned with land uses such as mining, deep well drilling, or others which do not involve exploitation of the immediate surface.

2
The basic requirements of a terrain evaluation system

As has already been indicated, there are three basic requirements for a terrain evaluation system: (a) a method for dealing with requests for information from intending users; (b) a capability for acquiring, analysing, systematising, and storing data about terrain and its actual or potential uses; and (c) a method of retrieving data from storage and translating it into the form required by the land user.

Analysis of user requirements

The first essential is an analysis of the information requirements of all relevant organisations and individuals concerned with land use. The main groups involved are: earth scientists, civil and military engineers, agriculturists of all types, foresters, hydrologists, urban, rural, and recreational planners, meteorologists and archeologists. Not only are such specialists concerned with widely contrasting types of land use but also with wide differences of intensity within the same land use.

Nevertheless, though large, the variety of terrain properties affecting land use is not infinite. Because many of them are of concern to more than one type of user, the task is not so difficult as appears at first sight. Most users would agree over the importance of a central core of information about land, which would include such properties as slope, soil particle size distribution, and moisture regime, and each would require additional specific data of less general interest. Only agriculturists and foresters, for instance, would need detailed information on soil nutrient status, and only engineers would require it on some aspects of soil strength. Finally, there is a large but finite number of terrain properties which would only be of importance to single narrow specialisations.

Nor is the range of land use intensities infinite. Interest in terrain becomes less at very large and at very small scales, because at these extremes the natural character of the ground becomes a relatively unimportant factor in the general land use equation. Planning at national and regional levels is concerned with economic and social considerations of which terrain forms only a small part. Likewise, the erection of major roads or buildings, the planning of small intensive urban gardens, or the making of small excavations involve so much effort or economic input per unit area that terrain factors become relatively insignificant. Terrain evaluation capability must therefore be concerned primarily with intermediate levels of land use intensity where the interest in terrain per unit area is neither so slight that its characteristics become unimportant, nor so intensive that their control becomes a relatively insignificant proportion of total cost. Experience has shown that the intensity of land use at which terrain evaluation is most important and relevant is that represented by the agriculturist concerned with the choice of farming systems, the engineer concerned with construction of 'B' roads, the battalion commander concerned with the movement of about a dozen vehicles, or a similar degree of land use interest in other fields. Such a level has been defined as 'moderately extensive'.

The main specialist interests concerned with terrain evaluation, and their requirements, are:

1. *Academic earth science.* This includes geologists, geomorphologists, zoologists, botanists, ecologists and pedologists. Such specialists differ from others interested in terrain in that their requirements are wider ranging, more variable, less routine and predictable, and less oriented towards immediate practical applications. They are as much concerned with past processes as with future possibilities. These characteristics make it especially difficult to find a single terrain evaluation system of general value to all. It is necessary to consider them to some extent as incidental contributors to, and beneficiaries from, such a system.

2. *Agriculture.* This can be considered broadly to include all those specialists concerned with raising economic plants and animals, and thus comprehends agriculturists, foresters, pastoralists, horticulturists, planters, etc. These are mainly concerned with three properties of land apart from its location: soil fertility, made up of nutrient status, texture, moisture regime, the absence of soil and topographic limitations, and a host of related factors; soil manageability consisting of tilth, hardness, permeability, relief, gradient, and their determining conditions; and the nature of existing vegetation.

3. *Civil engineering.* This involves a number of operations requiring knowledge of terrain factors: preparatory excavations for buildings, roads, railways, airfields, dams, bridges, canals, drains; quarries for borrow, lime, building stone, and brick. It must also consider the suitability of ground for load bearing, which can be derived from the particle size distribution, compressibility and shear strength of the soil under different moisture conditions, and from its liability to erosion and flash flooding, which is a function of slope, permeability and surface roughness.

4. *Military activities.* These include many of the operations of civil engineers, though generally with an emphasis on the less permanent works. Military interest in terrain also focuses on such aspects as artillery lines of sight, suitability of ground for excavating trenches,

fortifications, holding tent pegs, laying minefields, accepting parachute drops, and sustaining the passage and repassage of troops and of both tracked and untracked vehicles. Considerations of vehicle passage are known collectively as 'going' or 'trafficability', and depend primarily on soil strength, stickiness, and the frequency of gradients exceeding certain critical figures. Vehicle designers require, in addition, information on the distribution and areal extent of terrain attributes that pose special design problems.

5. *Meteorology and climatology.* These are concerned with the effect of terrain on weather and climate. Slope, aspect, and the nature of the soil surface influence climate both directly through their effect on winds, insolation, fog, cloud, and rain and indirectly through the activities of vegetation.

6. *Hydrology.* This requires knowledge of terrain in a number of ways, especially those relating to surface and subsoil water in defined territorial areas such as river catchments. Specifically, it is concerned with runoff regimes and quantities, stream flow, infiltration, and ground water depths and movements with practical application to water supplies, and erosion and flood hazards.

7. *Urban and rural residential and recreational planning.* Terrain is an important determinant of landscape aesthetics and as such must be considered in all planning schemes. In general, in developed areas, the less valuable land is for agriculture the more valuable it is for residential and recreational purposes. The requirements for these two types of land use are sufficiently similar to bring them into frequent competition. While agriculture tends to favour flat fertile areas of little aesthetic charm, recreational and residential developments prefer the proximity of hilly or rocky areas covered with forest or moorland and containing rivers and lakes, or sea coasts with sandy beaches in so far as their development can be harmonised with cost limitations.

8. *Nature conservancy and wildlife planning.* These interests also place a high value on landscape aesthetics but have an additional interest in geological or ecological rarity. Habitats of unusual plant or animal species have an increasing scientific interest and educational value so that in places their main importance may be for demonstration purposes.

9. *Archaeology.* The archaeologist is concerned with past land uses, and especially with anomalies and modifications to the natural land surfaces by early man. These mainly take the form of quarries, buildings, tombs, monuments, and earthworks.

Analysis of terrain classification requirements

Mabbutt (1968) has pointed out that the problems inherent in all terrain classification fall into three main groups: those of complexity, extent, and association, to which a fourth, scale, might be added.

Every spot on the earth's surface has a multitude of varied but intricately interrelated attributes which make it unique and difficult to compare with any other. It is necessary to try to understand and simplify this complexity so that the characteristics of different places can be defined, described, and compared.

Land can only be considered in terms of geographical areas. Difficulties arise over the fact that some attributes, such as landform, are clearly visible, while others, such as substrate material, are not. Some attributes, such as certain geological outcrops, have clear, sharp, limits; others, like the catenary succession of soils, do not. Some, like slope, refer to an area of land; others, like soil strength, refer to a point only. Any definition of a part of the landscape must therefore comprehend its vast range of properties, visible and invisible, definite and vague, large-scale and small-scale.

Scale is closely related to extent. Even within the range of land use intensities relevant to terrain evaluation there are marked differences in scale of interest, and within a given area a more intensive and detailed survey will reveal natural terrain units which would be missed by a reconnaissance. This is again a reflection of the internal variability of all terrain.

Again quoting Mabbutt, the 'character of land cannot be understood fully in terms of local controls acting in isolation but is in part determined by relationships with adjoining areas' from which they receive and which they give runoff, groundwater, microclimatic influences, and sedimentation. In short, they are 'open' rather than 'closed' systems.

Reproducibility versus recognisability of terrain

Terrain evaluation aims at making predictions. This capacity is especially important in view of its orientation towards reconnaissance assessments rather than towards detailed procedures involving comprehensive sampling. Predictions may be long range, allowing correlation between land types in different settings; medium range, between examples of a land type in a single region; or short range, within a single occurrence of a land type.

In order to make any sort of prediction two sets of facts are needed: the definition, and if possible the quantification, of the type of terrain in the known locality, and its recognition in the unknown locality. It is only by achieving both these objectives that information about the first can be transferred to the second. Two types of capability are therefore needed: recognition of types of landscape and knowledge of the properties of the types recognised, i.e. recognisability and reproducibility (Beckett and Webster, 1965a).

The recognisability of a type of terrain can be defined as the proportion of it which can be recognised, without ground check, out of the total area it covers, with the tools and resources available. These generally consist of published literature, aerial photographs of normal quality and scale (generally between 1:10 000 and 1:80 000) and the cover of topographic, geological, or soil maps normally available. In developed regions topographic maps may be at as large scale as 1:10 000 or better and geological maps at larger than 1:100 000 but in developing countries 1:100 000 and 1:2M respectively are more typical. Soil maps are generally at scales equal to or smaller than geological maps, but cover a much smaller percentage of the earth's surface. It can also probably be assumed that some person with experience of the terrain and climate in question can be found for the work. The recognition characteristics of terrain are not necessarily the same as its definitive criteria. While terrain types are defined on the basis of forms and materials, they may be recognised on aerial photographs by such incidental features as vegetation patterns and photo tone.

Reproducibility is the extent to which the properties of the terrain in question can be exactly defined in terms of earth forms and materials. Prediction between known and unknown areas is valueless unless these meet a minimum standard of accuracy.

It is clear that reproducibility and recognisability are more or less in inverse relation. The more carefully we define types of terrain, the more difficult they will become to recognise and the more types there will be. It is therefore necessary to reach a compromise between the two considerations, giving roughly equal weight to each, so as not to define the types beyond the point at which they are generally recognisable with the tools normally available, especially aerial photographs.

Information management requirements

The management of data about terrain does not differ in any important respect from the management of data in general. Its most nearly unique feature is its spatial element, viz. that it is related to geographical position.

There are essentially five stages in the process: acquisition, coding, storage, decoding, and issue. Automation can be used to some extent in all, but only dominantly in the third.

1. *Acquisition* includes all aspects of field survey, sampling, and laboratory analysis of samples which provide information on terrain. It also includes the abstraction of data from published books, journals, maps, and photographs, and from unpublished and unwritten sources of all kinds. It involves overcoming the general problem of inaccessibility to information sources and lack of access to site, and is materially assisted by the recognition of analogies between known and unknown areas.

2. *Coding* involves the classification and translation of the data so acquired as far as possible into numerical form in such a way that it can be fed into a store.

3. *Storage* includes not only the mechanisms for classifying and filing information in the form of books, pamphlets, maps, photographs, cards, microfilm, and tape, but also their manipulation so that required

information can be retrieved quickly and easily.

4. *Decoding* is the actual retrieval process itself and translates the stored information back into simple, manageable, and translateable form.

5. *Issue* is the final stage whereby the information is provided for the enquirer, and its format is governed by the requirement. Normally it is in the form of written matter, maps, annotated aerial photographs, and bibliographical references.

Not all stages will be carried out by the same people or organisations. Stage 1 is essentially a survey and research operation and is carried out by an organisation of this type. Stage 3 is straight data processing and is suited to specialists in this field, while Stage 5 requires editing, communications, and public relations skills. Stages 2 and 4 are intermediate and done by data processing specialists in consultation with surveyors and public relations men.

Administrative requirements

The management of a terrain intelligence system requires an administrative structure. The preceding section has outlined the specialist technical skills involved, but there are three aspects in particular which require further expertise.

First, general supervision is required of the whole process to ensure that the data input is relevant to the required output and to coordinate the stages in the process. Secondly, liaison must be maintained with those who provide and those who use the terrain data to keep pace with the changing needs of the latter and the changing capabilities of the former. Liaison must also be maintained with governmental, educational, and commercial organisations to publicise the contribution that terrain intelligence can make to their respective fields and to keep them informed of its developments. Thirdly, management is required of all the financial aspects: payment of staff, purchase of equipment, overheads, and income received.

3
The nature of terrain and its interpretation from maps

Terrain is the expression of the geological character, the soil, and the surface geometry of the earth's crust, and its general nature can be most easily understood by considering the maps which reveal these three aspects.

The geological basis[1]

Below a thin mantle of superficial material and soil, the earth's outermost shell is made up of rocks, which are in turn aggregates of minerals. They differ from one another in appearance and other properties depending on the type, relative abundance, grain size, and the mutual associations of the minerals present. The varieties of rock are almost endless, but if classified according to mode of origin, they fall into three major classes: *igneous, sedimentary,* and *metamorphic.* Unconsolidated superficial materials constitute a fourth class which may be called *drift.*

Igneous rocks are all those formed at high temperatures by solidification from a molten state. They can be either intrusive or extrusive. The former occur either as (1) large intrusive masses like lopoliths, laccoliths or batholiths, (2) small wall-like dikes, (3) cylindrical plugs, (4) sills sometimes thickening into phacolites. The first three types are intruded across bedding planes, the fourth along them. Extrusive rocks include volcanoes, lava flows, ashes, and agglomerates and occur as irregular layers or lenticles. All these types are shown diagrammatically in Fig. 3.1.

Sedimentary rocks are those deposited as sediments by ice, water, or wind either on land or more usually on the sea floor and subsequently consolidated. They generally occur in layers, known as bedding planes, whose relative age is determined by their location in the order of superposition.

The earth's surface is divisible into areas in which denudation is occurring to destroy the rocks at the surface, and areas upon which new material is being deposited. In any locality the geological history will have consisted at one time of denudation and at another of deposition.

1. For this section, the author is indebted to: A. A. Miller, (1965) *The Skin of the Earth.*

This alternation of conditions is reflected in the strata being separated by *unconformities* where they do not follow in direct sequence. These unconformities form the basis of the subdivision of geological time into the well-known stratigraphic subdivision from Pre-Cambrian to Recent. Where horizontal strata overlie others of strongly contrasting lithology and inclination, they are known as *overlapping*.

Most rocks undergo some tilting or folding after their deposition so that bedding planes no longer lie horizontal but have an inclination or *dip*. This may be defined as the direction of steepest slope on a bedding plane, and its direction is shown on geological maps by an arrow. The line drawn at right angles to the dip is horizontal and thus a contour along the bedding plane. Its direction is known as the *strike*.

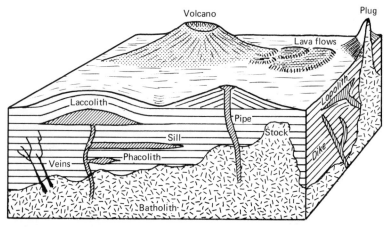

3.1 Igneous landforms

The area over which a bed of rock comes to the surface of the ground is known as its *outcrop*. This may be of almost any size or shape, depending on the slope of the ground and the dip and thickness of the beds. It can either be exposed or covered by a mantle of overlying drift. Extensive outcrops reflect structures caused by tectonic activity, the chief of which are folds and faults. The main types of structures caused by these are shown in Figs 3.2 to 3.19.

Folded forms:

Figure

3.2 A monoclinal fold of dipping strata giving simple strike ridges, cuestas, and vales.

3.3 A symmetrical syncline

3.4 A symmetrical anticline

3.5 An asymmetrical syncline

3.6 An asymmetrical anticline

3.7 Simple strike cuestas with related overfold

3.8 A pitching syncline and anticline
3.9 A dome with denuded crest
3.10 A synclinal basin
Strike faults:
3.11 Normal (hade[1] to downthrow side) with hade opposite to dip
3.12 As for Fig. 3.11, but hade with, but steeper than, dip
3.13 As for Fig. 3.11, but with hade less steep than dip
3.14 Reversed fault with hade opposite to dip
3.15 As for Fig. 3.14, but hade with, but steeper than, dip
3.16 As for Fig. 3.14, but hade less steep than dip
Dip faults:
3.17 In uniformly dipping strata
3.18 In an anticline
3.19 In a syncline

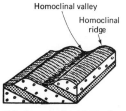

Homoclinal valley

Homoclinal
ridge

3.2 A monoclinal fold of
simple strike ridges,
cuestas, and vales

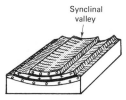

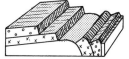

Synclinal
valley

3.3 A symmetrical syncline

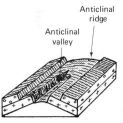

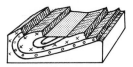

Anticlinal
ridge

Anticlinal
valley

3.4 A symmetrical
anticline

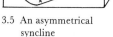

3.5 An asymmetrical
syncline

3.6 An asymmetrical
anticline

3.7 Simple strike cuestas
with related overfolds

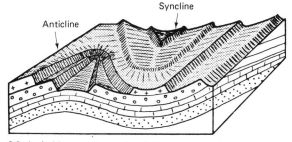

Syncline

Anticline

3.8 A pitching syncline and anticline

1. Angular deviation from the vertical of the fault line.

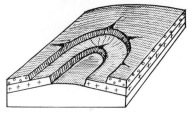

3.9 A dome with denuded crest

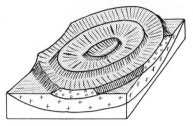

3.10 A synclinal basin

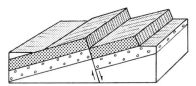

3.11 Normal strike fault with hade opposite to dip

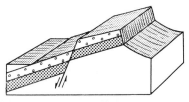

3.12 Normal strike fault with hade with dip but steeper

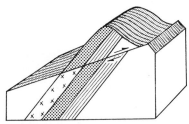

3.13 Normal strike fault with hade with dip but less steep

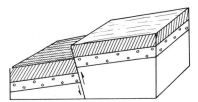

3.14 Reversed strike fault with hade opposite to dip

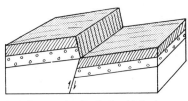

3.15 Reversed strike fault with hade with dip but steeper

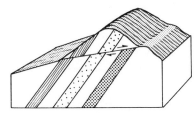

3.16 Reversed strike fault with hade with dip but less steep

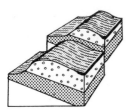 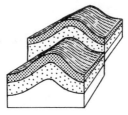 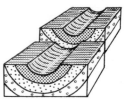

3.17 Dip fault in uniformly   3.18 Dip fault in an   3.19 Dip fault in a syncline
dipping strata   anticline

Metamorphic rocks are igneous or sedimentary rocks altered by the pressure and temperatures which accompany mountain building, and are generally harder and more contorted than the original types. Regions of metamorphic rock normally show a strong grain or *lineation* in the topography, though neither the ridges nor the valleys have the sharpness of folded sedimentary strata, and they are usually much fractured and faulted.

Drift deposits constitute the *regolith* and are always at the surface and rest unconformably on solid rock. They owe their form to their manner of deposition, which can be glacial, glaciofluvial, colluvial, alluvial, eolian, lacustrine, estuarine, or marine. They are composed of unconsolidated materials of all types, with particle sizes ranging from boulders to clay in mixtures of all proportions, depending on climate, parent material, and geomorphic history.

Generally, polar and desert climates give rise to dominantly mechanical weathering and yield coarser, more soluble, but less chemically altered fragments. In humid climates, chemical weathering is relatively more important, especially as temperature increases, and fragments tend to be finer, less soluble, but more chemically inert. The ratio of silica (SiO_2) to sesquioxides (Fe_2O_3 and Al_2O_3) tends to decrease with increasing temperature.

Rocks can be classified from a geomorphological point of view by their resistance to weathering and erosion and the type of materials they generally yield. Resistance is directly related to the permeability and the degree of consolidation of a rock, while the nature of its disintegration products depends largely on the proportion of siliceous material in the original rock. Siliceous rocks such as quartzites, sandstones, and flint yield coarser materials such as gravel, shingle, and sand. The higher the proportion of chemically basic material in the parent rock, the more vulnerable it tends to be to chemical weathering and the finer the texture of the drift it yields. At the extreme of fineness, mudstone, shales, and marls yield a high proportion of clay.

Soil

Soil is a natural body distinct from, yet transitional to, the parent material, and can be defined as the upper and biochemically weathered portion of the earth's surface. Its formation is directly or indirectly biological in nature, and involves more than the mere physical and chemical weathering of rocks.

The classification of soil for mapping purposes has always been quite distinct from the classification of terrain, though recently there has been an increasing convergence and harmony between the two, largely as a result of the use of aerial photographs.

The development of soil classification has really followed three lines: (1) *ad hoc* local classifications to meet immediate practical needs; (2) genetic schemes attempting to explain the worldwide distribution of major classes, and (3) attempts to fit both into comprehensive general schemes with a framework of quantitative definition. Although the three lines have to some extent developed together, they nevertheless broadly represent three stages in the overall development of soil classification.

Early soil classifications created classes using either rule of thumb local terms such as 'loam' or 'limon' or were based on single parameters such as clay or humus content. Such artificial systems are still widely used today at local level. The first recognition that soils were independent natural bodies with distinctive morphologies based on the effects of local and zonal soil forming agents was by Dokuchaiev and his school (Afanasiev, 1927) in the late nineteenth century. This made possible a 'natural' classification of soils, which has formed the basis for the development of the science to this day. The accumulating mass of data on soils since World War II has led to a third phase in which there has been progressive systematisation and qualification of soil classes based on natural morphological criteria. An outstanding example of this trend is the United States Department of Agriculture's 7th Approximation (Soil Survey Staff, 1960) which presents a hierarchy of exactly defined mutually exclusive soil classes whose definitive criteria aim at being increasingly morphological as one ascends the categorical sequence series: Series – Family – Subgroup – Great group – Suborder – Order.

Surface geometry

The third main aspect of terrain is the geometrical form of the ground. Detailed analysis and classification of this form will be left to later chapters, but its general character should here be outlined.

Landforms, being three-dimensional, can be understood from block diagrams, but detailed consideration is helped by excluding one of the dimensions and viewing them either in plan or in section.

Simply stated, landscape consists of hilltops, slopes, and valleys. Hilltops vary in locational arrangement and in profile view. The former can be seen from the 'hilltop envelopes' formed by map contours and the latter from crest-line diagrams.[1] In plan, crests can be either rounded

1. Methods of representing these are described by A. A. Miller (1965).

or elongated and parallel or randomly distributed, giving rise to the four possible distribution patterns shown on Fig. 10.2. The two sectional viewpoints contribute the other dimension, showing hills as peaks, ridges, or plateaux, each of which can occupy varying proportions of the total landscape. A statistical count of the number of hilltops in each altitude range may reveal a preponderant frequency at certain levels indicating, for instance, raised horizontal land surfaces remaining from an earlier age.

Slopes vary in length and steepness and can be convex, concave, or linear. Frequently, all three conditions can occur between the same crest and valley, the boundary between the two types being a *change of slope* if little marked, or a *break of slope* if strongly so. They can be assessed from maps from the spacing and distribution of contours and can be translated into profiles by graphical methods. Such profiles can be evaluated by other diagrammatic methods such as *area-height curves* and *hypsometric curves*. The former are graphs showing the proportion of land in a given area at each altitude, and therefore the relative amounts of highland, lowland, or land at any stated intermediate altitude. The latter are similar to these but are cumulative, expressing the total area above each altitude. Their value is likewise in detecting and demonstrating former land surfaces not so immediately obvious on maps alone.

There are three ways in which valleys can be viewed two-dimensionally. In plan they appear as a hydrographic network, while the two profile views are best taken as their longitudinal and transverse sections. Drainage catchments are occupied by hydrographic networks of branching streams in hierarchial order from smallest tributary to main channel. This hierarchy recognises a number of 'orders', first proposed by Horton (1945) and more recently simplified by Strahler (1964). Certain descriptive ratios to relate the orders have been evolved, the most important of which are *bifurcation ratio* and *stream length ratio*, discussed more fully in chapter 13.

Individual channels can be *straight*, *sinuous*, or *meandering*, depending on the degree of sinuosity of their courses, and *braided* if interrupted by numerous islands. The change from meandering to braiding occurs when the silt load of the stream exceeds a certain critical ratio to its velocity. Although many classifications of drainage patterns have been made, summaries such as those by the US Army Engineers Waterways Experiment Station (1963b), and Haggett and Chorley (1969) have simplified them and reduced them to two main groups: *integrated* and *non-integrated*, depending on the frequency of basins, lakes, swamps, and marshes. The main integrated patterns are *dendritic, rectangular, trellis, pinnate, parallel*, and *reticular*, as illustrated on Fig. 3.20, each of which can have a number of different forms and occur in many combinations. Non-integrated patterns usually occur on plains of glacial or river alluvium where drainage is poor or on limestone or evaporite rocks where it is partially subterranean. Some are shown on Fig. 3.21.

19

Dendritic: the commonest pattern. Indicates uniform materials.

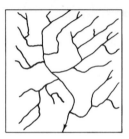

Rectangular: implies strong bedrock jointing and thin soil cover. The stronger the pattern, the thinner the soil.

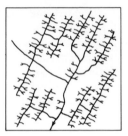

Trellis: implies strike ridge topography.

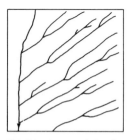

Parallel: characteristic of outwash areas of low topography, where main stream may indicate a fault.

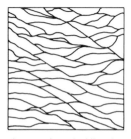

Anastomosing or braiding: in alluvial areas where sediment load exceeds carrying capacity of a stream.

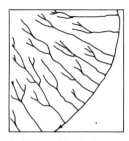

Radial (centrifugal): in isolated circular hill masses.

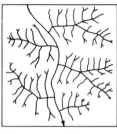

Pinnate: generally indicates high silt content as in loess or on flood plains.

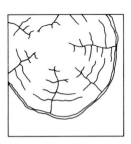

Annular: indicates igneous or sedimentary domes with concentric fractures or escarpments.

3.20 Types of integrated drainage pattern (after Way, 1968, pp. 6, 7)

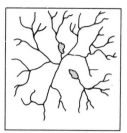

Deranged: with many ponds, bogs, or lakes. Indicates flattish landscape often glaciated.

Centripetal: a variation of the radial pattern with drainage towards a central point, usually a sink or the centre of an eroded anticline or syncline.

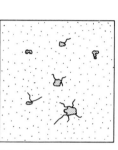

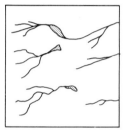

Internal: indicates highly porous level materials or karst conditions.

Dislocated: due to interruptions of drainage lines by faults or extrusions.

3.21 Types of non-integrated drainage pattern

The longitudinal profile or *thalweg* is taken along the winding line of the valley floor and is generally paraboloid and concave upwards with slopes approaching the base level of erosion asymptotically, as shown on Fig. 3.22. Most streams tend towards a state of *grade* in which there is a balance between erosion and deposition. Mackin (1948) established that such a graded condition represented an equilibrium whose diag-

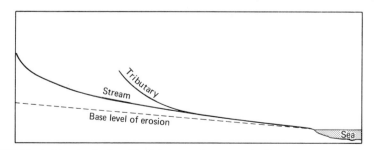

3.22 Typical longitudinal profile of thalwegs of stream and tributary

nostic characteristic was its tendency to absorb the effects of any displacement caused by changes in the controlling factors. Grade can be interrupted by convexities or breaks of slope in the stream bed. These may be due to the outcrop of resistant rocks, to glaciation leaving parts of valleys hanging above their lower reaches, to rejuvenation at some intermediate point in the valley, or to flattening due to the upstream encroachment of alluviation.

Valley cross-profiles are variable and often indicative of geomorphic origin. There are a number of distinct types:

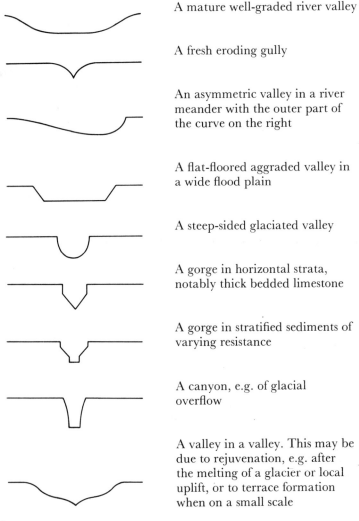

A mature well-graded river valley

A fresh eroding gully

An asymmetric valley in a river meander with the outer part of the curve on the right

A flat-floored aggraded valley in a wide flood plain

A steep-sided glaciated valley

A gorge in horizontal strata, notably thick bedded limestone

A gorge in stratified sediments of varying resistance

A canyon, e.g. of glacial overflow

A valley in a valley. This may be due to rejuvenation, e.g. after the melting of a glacier or local uplift, or to terrace formation when on a small scale

4
Principles of regional and terrain classification

Terrain classification is the first stage of the data acquisition process and the one that most concerns the earth scientist. Its basic concepts are derived from regional science, so this must next be considered.

Principles of regionalisation[1]

Regionalisations of the earth's surface, although mainly the province of geography, have never been confined to it. Among the sciences they have been undertaken in ecology, soil science, and climatology, while among the social sciences they have been used in cultural anthropology, urban and rural sociology, and economics (e.g. by Isard and his school).

Within geography, the regional concept has had a long and complex development, but has generally combined physical and human factors into general schemes. In the eighteenth century, political units were found to be inadequate and more 'natural' groupings were sought in their place. This led to the monographs on the *pays* of France, distinguished by their characteristic *genres de vie*. In 1905 Herbertson subdivided the whole world into *natural regions* based on climate, configuration, and vegetation. He was followed by Dokuchaiev, who suggested a soil scheme for the world based on climatic-vegetational zones.

The difficulty of achieving a general classification into natural regions and the recognition that its basis should be the things regionalized rather than an assumed cause led to a new approach. Workers in different sciences sought schemes of simpler and less fundamental *single feature regions*. Notable are those of de Candolle (1856) and Schimper (1903) for plants, Wallace (1876) for animals, and Köppen (1931) for climate.

These three trends: the pays concept, the natural region, and the single feature region, were further developed between the wars. The last, in particular, was followed by Thornthwaite's refinement of Köppen's climatic classification (1933) and the attempts by Unstead

1. For this section the author is indebted to D. Grigg (1967), 'Regions, models, and classes', in Chorley and Haggett, *Models in Geography*, pp. 461–509.

(1933) and Whittlesey (1936) to regionalise cultural and agricultural factors respectively.

Two further advances in the regional concept date from this period. A committee of the Geographical Association (Unstead *et al.* 1937) distinguished *specific* and *generic regions*, the former being unique and the latter being recurrent and enabling analogies to be drawn between different areas. For example, 'the Chilterns' are a specific region, but 'chalk cuesta' is a generic term which could relate to other regions besides the Chilterns.

Again, Dickinson (1930) and Christaller (1933) evolved the idea of the *nodal region*, based on interconnections between a central place and its surrounding countryside. This has come to denote regions which are defined by a functional organisation of unlike properties, by contrast with *uniform regions*, which represent a uniformity of properties over a geographical area. Although primarily designed for urban geography, the concept has application in the physical field to regions for which there is a diffusion of, say sediment or plant species from a central source.

Regions are usually studied at a single point in time. It is important, however, to recognise that they are not static but are continually undergoing change. This is especially true of the internal organisation of nodal regions where the node itself has a changing role. For example, an isolated anticline composed of concentric layers of different rock will yield a changing pattern of sedimentation as its inner layers are exposed by erosion.

These developments in the regional concept have had the effect of re-emphasising the old problems of complexity, extent, scale, and association previously noted, and of introducing new ones. The difficulty of delimiting regions accurately calls the whole reality of their existence into question.

The debate as to whether or not regions are concrete objects finds analogies in other disciplines. In history or sociology the question arises whether communities or nations really exist as organisms or are just collections of individuals. In ecology, the same question must be asked of climax associations, and in soil science of soil groups. In essence, the debate is that between *realism* and *nominalism*, first discussed by Aristotle, and put in the context of soil science by Robinson (1949). The realist maintains that abstractions like 'table', 'region', 'community', 'podsol', etc. have a real existence of which the individual examples are expressions. These may deviate more or less from the ideal, but the ideal would still exist in the absence of any example of it. The nominalist maintains, to the contrary, that abstractions are mere names, and have no existence apart from the examples which compose them. There is thus no such thing as an abstract 'table', but 'table' is merely the name we arbitrarily assign to a group of objects having certain recognisable characteristics in common. This controversy has not been resolved, but is mentioned here because it finds its echo in presentday regional science. Some, including notably the Russians, claim that there are absolute physical and human regions and that it is the aim of science to discover and elucidate them. It is hard, however, to reconcile this with the difficulty of defining particular regions and more specifically with the concept of nodal regions which grade imperceptibly into one another. On the whole, the nominalist view is the one which seems to have the fewest difficulties and is adopted in this book. In consequence, an arbitrary element is accepted as inevitable in all regional delimitation.

Whatever the view taken of regions, boundary fixing remains difficult. One solution is to consider each region as having a core area surrounded by transitional zones to other regions, which may be larger than the cores. The cores are analogous to *culture centres* in anthropology, or *formations* in plant geography. These assume the existence of transitional areas, which in the latter science are called *ecotones*. Poore (1956) has suggested the word *node* for the cores, and has emphasised that these may be smaller than their ecotones.

When boundaries are to be drawn they must be based on a number of criteria. The greater their number, the less likely their coincidence, and the harder the task of harmonising them into a single boundary line. Grigg (1967) discusses the methods used for resolving this problem, of which there are basically two. *Maull's girdle method* proceeds by drawing maps of the regional limits of each land property in question, superimposing these on a single map, and then selecting the line where more than a given number of boundaries approximately coincide. The method can be refined by drawing in a calculated mean boundary line, or by incorporating calculations of the sizes of overlapping areas multiplied by the number of properties involved in the overlap. But this method becomes increasingly complicated and unsatisfactory as the number of criteria is enlarged. The second method is the statistical *principal components* method suggested by Hagood *et al.* in 1941, and the associated technique of direct factor analysis. In these methods the correlation between a large number of properties is measured and the significant criteria extracted from the initial properties considered.

The philosophic problems of regionalisation would be eased if regions could be considered as classes analogous to those used in other sciences, because this would much simplify information management and the use of quantitative techniques. Grigg (1967) has pointed out that in several respects regions can be so regarded. They can be arrived at either deductively by logical subdivision of a global population or inductively by the classification of numerous local examples. To this extent, regionalisation is akin to classification in the methods used. The distinction between uniform and nodal regions also parallels the classification systems of other sciences where classes may be based either on similarities among the objects contained in them or on the interrelationships between such objects when these are themselves dissimilar. Concluding that regions are areal classes, we can roughly equate places with individuals, areas with classes, and boundaries with class limits. The analogy breaks down mainly over the fact that every region has one unique feature which defies classification – its location – but with this last caveat, it is justifiable to treat regions as scientific classes.

Classification systems for terrain

In classifying terrain for practical ends, two alternative approaches are possible.

First, one can consider the terrain itself, classifying it into natural units and then attempting to measure their properties quantitatively and to relate them to land uses.

Alternatively, one can consider the terrain from the point of view of the uses envisaged and devise a list of the relevant land attributes and the class limits required within each, and then map each one. A superimposition of these maps will then give a complete classification. As an illustration, one can say that soils in Britain become more suitable for the cultivation of wheat with decreasing altitude and gradient and increasing clay content. If one then classifies each of these parameters into acceptable and unacceptable classes, one can map an area of landscape for each of them separately, overlay the maps to obtain another showing all the combinations of classes, and then grade the favourability of its various parts in accordance with their showing in the three parameters. As the number of land uses or parameters enlarges, for example if one also wants to know about barley and include a consideration of soil moisture or temperature, the operation becomes more complex.

These two approaches have important differences. The first, which has come to be called the 'physiographic' or 'landscape' method, will be favoured to the extent that the number of land uses is increased, and where resources are limited or access to site is difficult. If information is required over the whole range of practical needs in agriculture, forestry, engineering, mining, vehicle mobility etc., the number of significant criteria becomes too large for it to be practicable to measure them all or to overlay parametric diagrams of them. This leads inevitably to a general reliance on a physiographic classification of the landscape itself, encouraging extrapolations to be made between analogous areas.

The second method, which has come to be called the 'parametric method' tends to emphasise precise measurements and is most useful where only a few land uses are being considered and computing facilities are available. Because of the dependence on detailed measurements and the difficulty of extrapolating these into unknown areas by means of physiographic analogy, it tends to be better suited to the detailed analysis of small areas than to broad reconnaissance.

5
Genetic and landscape systems: general principles

The basis of genetic and landscape classifications

As Beckett (1962) points out, we may start from the commonplace that every country may be divided into a finite number of physical regions, each with a characteristic landscape. This is recognised in lay parlance by the use of such terms as Cotswolds or Chilterns in Britain, Champagne or Jura in France, Great Plains or Piedmont in USA, etc. Such regions are clearly distinct from their neighbours and usually recognisable both on the ground and from the air.

The unity of such regions can be considered as due to a common tectonic and climatic history acting on a broadly uniform rock. Comparable climates acting upon similar rocks in regions of like geomorphological history in different parts of the world tend to give similar landscapes. It is the basic assumption of both genetic and landscape systems of terrain evaluation that this gives rise to analogies between landscapes in different parts of the world which are sufficiently close for it to be possible to make predictions from the known to the unknown.

As the objectives of terrain classifications are practical, it is inevitable that the selection of natural criteria in defining landscape classes be biased towards practical requirements.

The most useful type of classification, as pointed out by J. S. Mill (1891), is that whose classes permit the widest range of generalisations to be made. For this reason, classifications of natural phenomena should be genetic and select criteria which are the causes of other things, or at least sure marks of them. In general, therefore, the best type of terrain classification is that which separates the landscape into natural units based on origin, process, and form. Such units represent the result of the interaction of all the genetic factors in giving a single ecosystem and have the practical advantage of integrating the complex interrelationships of the many attributes of the landscape into a single whole.

Secondly, the criteria used in defining the units should be chosen from the fundamental and permanent characteristics of the landscape, and avoid not only those more loosely related to it, such as climatic factors or the works of man, but also properties which are somewhat ephemeral, such as flora, fauna, or ecology. These are better considered as attached information than as definitive of classes.

Thirdly, landscape units must generally be regarded as *uniform* rather than as *nodal* regions, in that their internal homogeneity consists in a uniformity of natural properties rather than an organisation of unlike functions. This is because uniform regions generally have clearer and more easily defined boundaries and clearer practical applications to the natural landscape than do nodal regions. Regions based on rock types and landforms, for instance, are generally more useful than those based on the diffusion of some form of sediment or chemical from a source. The 'unity' implied in the term 'uniformity' is not absolute, but inevitably conditioned by the scale of mapping and the subjective interpretations of the compiler.

The natural criteria must be balanced by the overriding practical need for simplicity. This is achieved by keeping the number of landscape classes to a reasonable minimum while at the same time ensuring that each is a valuable vehicle for data storage, i.e. that it is internally homogeneous and distinct from others.

Practical advantages of landscape systems

Landscape systems have three outstanding advantages over parametric. They help explain the fundamental causes of landscape differentiation; they assist reconnaissance; and they facilitate the appreciation of regions as a whole.

Viewing landscape as a whole in order to determine its constituent units inevitably leads to a consideration of origins, processes and genetic relationships. This has value in leading to an understanding of imponderable factors such as the effects of climate and tectonic activity and also in permitting predictions of natural hazards such as erosion and flooding, and the occurrence of exploitable land resources. Veins and fluvial diffusions of certain economic minerals occur in patterns which are related to certain landforms, and springs can be inferred from particular combinations of rock outcrop and position.

An integrated view of landscape is especially valuable in reconnaissance surveys because it unites all aspects and types of practical data on to the same geomorphic base. This brings together the interests of all land users concerned with the top six metres of the earth's crust by providing them with background information, and saving them costly duplication of field effort. Even in areas where a substantial amount of information about resources already exists but needs reassessment for land use planning, there are advantages in groups of specialists undertaking it jointly and on a common conceptual basis.

Thirdly, regions can be more easily appreciated as a whole when considered on a landscape basis. It eases the task of delimiting areas with different kinds of potential and demanding different types and priorities of effort. It is often more important to know the existence, location, and size of such areas and how they are related to one another geographically and to concentrations of population, transport systems, and markets, than to have a very detailed idea of a single resource.

Large scale landscape units: the genetic approach

Large units have been developed from deductive classifications by logical subdivision of landscape on the basis of causal environmental factors. Among the first to use this approach was Herbertson, who in 1905 based his natural regions primarily on climate and secondarily on morphology. Fenneman, in 1928, followed Davis's emphasis on the genetic development of landforms by basing his physiographic provinces of the USA on morphogenesis.

The broad genetic classifications of landscape which use these large units have, as Mabbutt (1968) points out, three practical advantages. First, they provide a coordinating framework for smaller units derived by other methods of classification, which is useful for correlating analogous areas. Secondly, they aid in the reconnaissance of large areas by providing an overall view of genetic relations which helps economize effort in field sampling. Thirdly, they have educational value in explaining world patterns, and are especially valuable to specialists such as historians and planners.

On the other hand, they have a number of disadvantages which restrict their practical usefulness. Because the units are large, usually extending to thousands of square kilometres, they have both great internal complexity and vague boundaries. The lower size limits to practicable genetic subdivision cannot be reduced to give units small enough to be homogeneous for most types of land use. Precision of definition is sacrificed in exchange for an overall correlative framework.

This vagueness of definition in large units caused an early search for smaller and more precise ones. Heath (1956), for instance, pointed out how Bowman in his studies of Andean settlement (1916) could only use such divisions as a general background and required greater detail for the study of actual land use. The Michigan Economic Land Survey of 1922 quoted by Mabbutt as perhaps the earliest systematic attempt at regional land evaluation, also required detailed definition of land units, and it is the smaller of these which have become the basis of the landscape system.

Small scale land units: the landscape approach

Landscape methods using smaller land units have been developed by a number of organisations in several countries, pursuing different objectives. Most countries have seen some development along these lines. They have generally aimed at data acquisition for agricultural, engineering, military, or planning purposes, and are considered in more

29

detail in chapters 9–14.

Almost all landscape methods have a number of characteristics in common. There is general agreement not only in recognising a hierarchy of land units of different sizes but also in according priority to those at two levels of magnitude smaller basic units, hereinafter generally referred to by the term *facet*, which are homogeneous for practical purposes and suited to a mapping scale of 1:50 000–1:100 000, and a larger grouping unit, hereinafter generally called a *land system*, suited to mapping scales of 1:250 000–1:1M.

Units of land system size are generally considered to be recurring. This is important in opening up the possibility of recognising analogies between similar, but separated areas. It also distinguishes them from the larger genetic units which are regarded as non-recurring.

Facets have a character which is part natural and part artificial, depending on the tools used in their recognition. The more exact the definition and the more homogeneous the properties required of them, the harder it becomes to recognise them on maps or aerial photographs, and on the ground with the tools currently available. This necessitates a compromise by which such requirements are not pressed beyond the point at which the facets cease to be generally recognisable.

Distinction is made between the criteria by which facets are defined, and which are few, fundamental, invariable, and always present, and those by which they are recognised, which are neither fundamental nor necessarily always the same. Vegetation, for instance, is generally the most useful clue to facets on aerial photographs, but is seldom definitive.

Some writers appear to regard the difference between these units as only a matter of scale, e.g. Christian, (1958), and Vinogradov *et al.*, (1962). Others such as Beckett and Webster (1962) recognise a fundamental difference in kind that land systems are basically morphogenetic, and facets are basically physiographic units. Mabbutt (1968) appears to favour the latter view, but for the reason that facets are characterised by an unbroken continuity of internal properties that land systems cannot have.

6
Parametric systems of terrain evaluation: general principles[1]

Parametric land evaluation can be defined as the classification and subdivision of land on the basis of selected attribute values. The simplest form of parametric map is one which divides a single factor into classes at certain critical values to give a simple isarithmic map, i.e. as the contours round chosen altitude classes give a hypsometric map. Additional quantitative parameters, such as isohyets or isotherms, can be chosen and contoured in the same way and then superimposed to give increasingly sophisticated and complex maps. Alternatively, numerical criteria can be added to qualitative or descriptive maps to give them a more quantitative framework.

The general parametric method is demonstrated graphically in Figs 6.1–6.4 by showing how single terrain attributes can be combined to give a composite parametric map. Fig. 6.1 shows hypsometry, Fig. 6.2 rainfall, and Fig. 6.3 geology of south-eastern England. Fig. 6.4 is a composite parametric map of a simple type which shows the combination of the three attributes on a single sheet.

Although parametric classifications are old, they have only been developed for terrain evaluation in recent years parallel to, but largely independently of, landscape systems. The method has been most fully developed for military purposes. The chief agency has been the United States Army Engineer Waterways Experiment Station at Vicksburg, Mississippi (1959) hereinafter referred to as USAEWES, but different aspects of the same general quantitative parametric approach can be seen in other US military research such as that by the Quartermaster Research and Engineering Centre at Natick, Mass. (Wood and Snell, 1959), the Air Force Cambridge Research Laboratories (e.g. Ta Liang, 1964), the Office of Naval Research (Melton, 1958), and Cornell Aeronautical Laboratory Inc. (1963). It is also used in military land evaluation by the Canadian Army (Parry *et al.*, 1968); for agricultural land evaluation in Canada and Britain (Bibby and Mackney, (1969), in USA (US Department of the Interior, 1951), and in Russia (Ignatyev, 1968); and for urban site and recreational planning, especially in USA (e.g. Kiefer, 1967).

1. In this chapter the author is indebted to: J. A. Mabbutt (1968). 'Review of concepts of land classification' in Stewart, *Land Evaluation*, pp. 11–28.

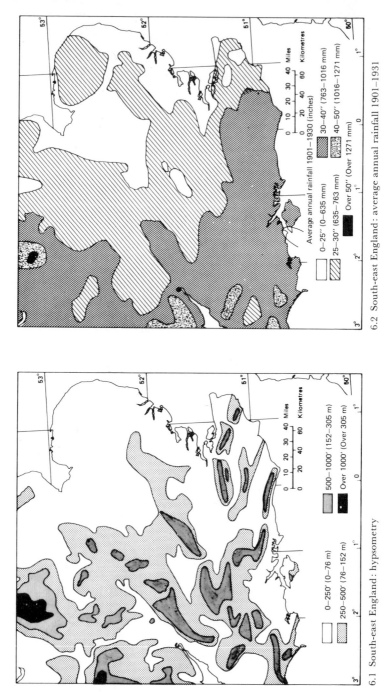

6.2 South-east England: average annual rainfall 1901–1931

Average annual rainfall 1901–1930 (inches)

0–25″ (0–635 mm)   30–40″ (763–1016 mm)

25–30″ (635–763 mm)   40–50″ (1016–1271 mm)

Over 50″ (Over 1271 mm)

6.1 South-east England: hypsometry

0–250′ (0–76 m)   500–1000′ (152–305 m)

250–500′ (76–152 m)   Over 1000′ (Over 305 m)

Miles
Kilometres

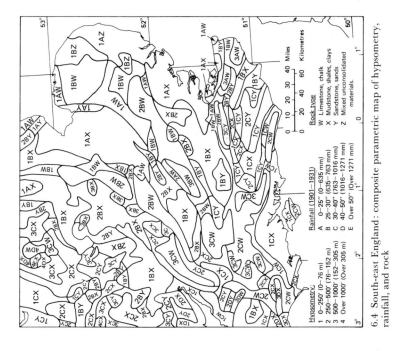

6.4 South-east England: composite parametric map of hypsometry, rainfall, and rock

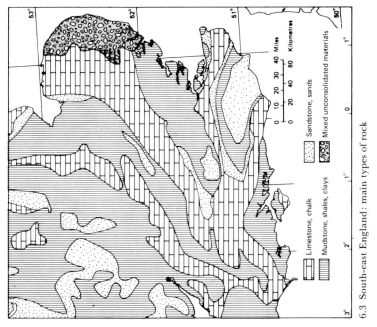

6.3 South-east England: main types of rock

When parametric classifications are used for terrain, certain inherent problems must be considered. The chief of these are the choice of attributes to be mapped, their subdivision into classes and the recognition of these classes on the ground. This is especially difficult because of their essentially arbitrary nature.

Choice of attributes and their subdivision

There are certain basic requirements for terrain attributes if the map is for a practical purpose. They must be relevant to the land use being considered. They must be recognisable and measurable in the field, and they must, like landscape units, define landscape at levels equivalent both to that of the land system and the facet. The requirements for these are somewhat different. At a level of generalisation analogous to that of the land system in the landscape method, parameters must either record the general character of an area in terms of a specific feature such as 'the most commonly occurring slope class' for which the mapping unit is taken to be homogeneous, or else must record the arrangement and mutual relations of components such as the distribution pattern of upland areas. This distinction is comparable to that between uniform and nodal regions in geography (Grigg, 1967).

At a level of generalisation equivalent to the facet the dominant need is to find parameters which define the form of the land including its shape, profile, and dimensions on both a meso- and a micro- scale as well as the earth materials and the moisture regime. Individual parametric units must be distinguished on the basis of the internal homogeneity or at least the internal predictability of these properties.

The separation of areas in a landscape system of classification is based on the occurrence of recognition features such as vegetation or rock form. With a parametric system it is necessary to fix limiting values. Such values are often determined from some land use criterion such as the crop requirement, for instance, for a certain minimum percentage of clay in the soil or a maximum to the exchangeable sodium percentage on the exchange complex. The problem is more difficult when such data are not known, and reliance must then be placed on the quantification of those natural breaks in the landscape which can be selected as significant directly or diagnostically to land use.

In the last resort it may be necessary to use totally arbitrary mathematical subdivisions of an attribute.

Depth and time variations within classes

It is possible to choose parameters, subdivide them into classes, and map them using the two surface dimensions out of the possible four in the landscape, i.e. considering only the lateral distribution of the surface cover and forgetting both depth and time. Soils and land forms reveal variations in underground profile and changes through time which cannot be comprehended on a two-dimensional map of the surface. Under the landscape approach these complexities are simplified by first compounding them into soil classes which are recognisable from causal or associated factors such as parent material, landforms, or vegetation visible at the surface. Under the parametric approach, however, which cannot make use of such associations and analogies, it is necessary to correct ignorance of the subsoil by surveying in greater detail, using more refined scanning techniques, spacing samples more closely, or economising effort by making a narrower selection of properties which experimentation shows to be most closely related to the land uses envisaged.

Moreover, while a landscape classification automatically separates areas with different seasonal or other temporal changes, a parametric classification must make special allowance for this. It is especially important to include such parameters when surveying areas with extreme seasonal changes of temperature or moisture, especially when the two occur in combination, as in circumpolar or monsoonal regions.

Mapping parametric classifications

A parametric field investigation yields, for each factor considered, an array of numerical values or quantative statements related to a grid of sample points. The next problem is to map them. This can be done by constructing a map on which contours known as isopleths are drawn round values above and below arbitrarily selected critical levels.

Alternatively, a *trend-surface* map can be drawn for each factor. Trend-surface mapping is a system whereby numerical data from a network of points on a two-dimensional surface is represented by means of an illustrative three-dimensional diagram 'contoured' according to the data from the points. The technique is summarised by Cole and King (1968). It gives a map which can be used towards defining land units or in the study of the attribute in question.

When the aim is to delimit land units, a number of procedures are possible. First, one can superimpose maps of different attributes to form a composite mosaic. Such a mosaic might, for instance, show all possible combinations of gradient, percentage clay, and pH of the soil. Frequently occurring combinations might then be recognised as landscape complexes and related to physiographic types. Secondly, one may begin with isarithmic maps of independent attributes. These can be superimposed on each other to outline areas by one of three methods: by using only the highest of the factors in each spot, by using approximately coincident boundaries from different overlays which lend themselves to simple adjustment, or by combining the information from attributes on different overlays which appear to be complimentary.

Advantages and disadvantages of the parametric system

It is of value to summarise some of the main advantages and disadvantages of the parametric when compared with the landscape method. First, its advantages. It is more quantitative and less dependent upon subjective interpretations of landforms. It is a more statistically reliable means of measuring variance, formulating rational sampling policy and expressing the probability limits of findings. It is better adapted to the use of the new remote sensors which are able to scan directly those terrain attributes which have hitherto had to be inferred from associated features.

These sensors include not only aerial photographs using the visible and near infrared parts of the spectrum, but also infrared linescan, side-looking radar, and such geophysical tools as the magnetometer, gravity meter and scintillation counter. The parametric approach is also better suited to the increasing use of electronic data handling which favours information in quantitative form. Finally, it leads to a system which is more flexible, giving land units which are more easily modified in the light of expanding knowledge.

There are also some disadvantages. It is difficult to decide on the right parameters and the class limits to measure for any given land use. With very few exceptions exact measurements of the earth's surface cannot be extrapolated beyond their place of measurement other than by reasoning from physiographic analogy between landforms recognised on aerial photographs. Because the predictive capacity of a parametric system is severely limited in this way, more detail is needed than in a landscape system. This necessitates more ground measurement and greater detail in mapping each parameter. As information on most attributes tends to be sparse, maps must either be based on slow and costly ground surveys which tend to be restricted to small areas, or else be at scales that are too small to be of much practical use to most people. Finally, because information is based on point samples only, accuracy tends to fall off rapidly between the points unless remote scanning of attributes has been possible to fill in the intermediate areas.

On the whole, the parametric approach sacrifices comprehensiveness and ease of recognition for the reliability and quantitative output of a definition based on measured properties. It has the advantage today that the picture it gives is becoming increasingly complete and readily obtained with the new techniques of scanning and computing becoming available.

Despite the differences between them, however, the landscape and parametric methods should be thought of as ultimately complementary rather than conflicting. The relative advantage of each varies with circumstances. With the standard equipment and expertise of today, the landscape approach offers the possibility of more rapid survey at lower cost. Its reliability is adequate for reconnaissance and, with moderately close sampling, for semi-detailed surveys, and has the advantage of combining all survey work into a single operation. For detailed survey, however, the greater precision and reliability of the parametric approach tend increasingly to outweigh the advantages of extrapolation to analogous areas and lower costs offered by the landscape method.

7
Vegetation in terrain evaluation

The place of vegetation in terrain evaluation

There is almost no part of the earth's land surface, except permanent snow and ice fields and the most extreme deserts, which does not support some vegetation. In humid climates the cover is generally thick enough completely to obscure the surface of the ground.

There are three ways in which vegetation is important in the evaluation of terrain: as an index for the recognition of terrain types, as an attribute in their definition, and as a natural resource physically attached to them.

Vegetation is an important key to the recognition of terrain types, particularly when they are viewed from the air. This is not only true in undeveloped areas of low population where there is a sort of natural ecological climax closely reflecting site characteristics, but also, although to a lesser extent, in developed areas such as western Europe where a close relationship has evolved between agricultural land use and site. The most important single factor in these relationships is probably the effect of ground configuration in determining soil moisture conditions at any given site. For this reason vegetation is an especially valuable index of terrain in areas where soil moisture variations cross limits which are critical to plant growth, as in sub-arid lands. Other terrain factors such as slope, aspect, soil depth and nutrient status are also important, and are each determinants of plant cover in some locality.

A good example of the use of vegetation as an index was the survey of the range potential of land in Jordan carried out by Hunting Technical Services Ltd. (1954) (Fig. 7.1). The CSIRO have also made especially wide use of it in recognising terrain types in their surveys of semi-arid Australia (Mabbutt and Stewart, 1965).

Although widely used in recognising terrain types, vegetation has not normally been regarded as definitive of units. There are two main reasons for this. First, the plant population of an area is an intrinsically different phenomenon from its terrain, and cannot logically be regarded as a part of it. The two can vary independently, so that changes or natural boundaries in the one are not necessarily reflected in the other. Secondly, vegetation is an ephemeral rather than a permanent charac-

RANGE TYPE 4v	10a	11
TOPOGRAPHY Wadis in the limestone and basalt desert region. Carry off rare water.	TOPOGRAPHY Wadis with deep central drainage channel.	TOPOGRAPHY Muddy depression occupying internal drainage areas or stretches of broad silty wadis.
VEGETATION Brush of various species, *Haloxylon salicornicum, Anabasis, Artemisia herba-alba, Phlomis* spp., etc.	VEGETATION Open fringing woodland and *Pistachia atlantica.* Ground vegetation of *Retama ractam, Atriplex, Halimus* and *Amygdalus spartioides.*	VEGETATION Usually nil.
PHOTO CHARACTERISTICS Light grey tone often much the same as surrounding areas. Drainage channels emphasised by vegetation showing as dark grains.	PHOTO CHARACTERISTICS Light grey with darker shading—individual trees showing up as dark spots.	PHOTO CHARACTERISTICS Generally white or very light grey smooth areas.

7.1 Sample sheet and range classification key showing some range types in Jordan (Source: *Report on the Range Classification Survey of the Hashemite Kingdom of Jordan*, 1956, p. 13. Jordan – United States joint Fund for Special Economic Assistance. Department for Range and Water Resources, by Hunting Technical Services Ltd.)

teristic of landscape. It is in a continuous state of change through seasonal variations, natural colonisation, the activities of wildlife, grazing, burning, plant disease, clearances and introductions for agricultural purposes, or other natural or human-inspired processes. There are exceptions to this general rule where, in some instances, vegetation units have been regarded as definitive, rather than merely as diagnostic, of terrain units. The earliest expressions of this view seem to have been from Bourne (1931) and Veatch (1933) who distinguished *sites* and *natural land types* on ecological as well as physical grounds. Hare (1959) relied even more strongly on vegetation units in the reconnaissance survey of Labrador-Ungava. The same viewpoint is still accepted by the Land Research and Regional Survey Division of CSIRO, who regard vegetation as well as landforms and soils as a definitive criterion for land systems (Mabbutt and Stewart, 1965). The reasons for this emphasis is clear. Vegetation, even in an unimproved state, provides the natural forest and range on which much of the world's population depends for timber, fuel and grazing, and over vast areas constitutes virtually the only natural resource. In such undeveloped areas as are considered in these surveys, its character often closely reflects the underlying physiography.

Approaches to the problem of vegetation description

There have been three basic approaches to the problem of viewing vegetation in its relation to terrain. First, there is the botanical method. This views plants in terms of their position in the classical Linnean groups and seeks to determine their geographical distribution. The applications of this approach are mainly botanical and relate to the colonisation of and succession in land areas by different plant groups. Its value to the study of terrain derives from the proof of relationships between particular plant species and types of terrain and it is usually best adapted to areas where physical rather than chemical conditions are the prime determinants of plant growth.

Two examples of the use of this general method are those by Smith (1949) in the Sudan and Beckett and Webster (1965d) in Britain. The main limitation on natural plant growth in the Sudan is almost invariably aridity. Smith showed that the natural occurrences of certain tree species could be accurately predicted if only three variables were known – rainfall, soil texture and site. Given the same clay soil and site type, a succession of different Acacia species occurred over the 1500 kilometre transect from the river Atbara at about 16°N to the Sobat near Nasir at 9°N. Conversely, a single species, such as *Acacia senegal*, occurred generally on sands under 400 mm of rainfall and on clays under 600 mm. So general was this relationship that it was possible to say that a tree species in the Sudan required 1·5 times as much rain on clay soils as they did on sands. It was further found that, with no recorded exceptions, all species progressed through their rainfall span via the same sequence of site types, and that, given the same rainfall, the site types would be arranged into a transect in order of their provision of soil moisture for tree growth. This site transect is represented diagrammatically in Fig. 7.2, and illustrates a useful relation between plant species and terrain types from which rough predictions of each can be made from the other. It is probable that similar relationships exist elsewhere.

In a study of similar type in the widely different Oxford area, Beckett and Webster (1965d) found statistically significant correlations between land facets and certain tree, shrub, and climbing species in hedgerows. On the whole, shrubs showed the closest degree of association. For example, willow and guelder rose were found where there was shallow ground water, crabapple on well-drained sites, and blackthorn in areas of highest fertility.

The second way of classifying vegetation in relation to terrain is by associative groups of plants, or ecologically. This is the normal method employed in atlases, textbooks, and on small scale maps generally. The terminology is half botanical and half physiognomic and uses such terms as 'needleleaf evergreen', 'grassland', 'Acacia tall grass scrub' etc. Although many studies have been made in this way, including such classics as Tansley's *The British Islands and Their Vegetation*, the primary aim has been to show the interrelationships of plants to each other and with the total environment rather than to emphasise their relationship to terrain.

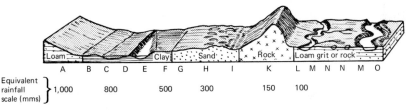

DESCRIPTION OF SITE TYPES

A. Hard-surface slopes subject to sheet flow.
B. High flood plain (flooded only for days).
C. Low flood plain (flooded for weeks at a time).
D. Mounds or banks in swamp or beside rivers.
E. Beds of land-locked pools.
F. Flat clay plains without run-off or standing water.
G. Sand plains.
H. Sand dunes and hills.
I. Pockets, hollows, or valley beds in sand country.
K. Rocky hill slopes.
L. Seasonal water courses flushing for an hour or two after rains.
M. Hard plains of grit or rock.
N. Seasonal runnels flushing for an hour or two after rains.
O. Banks of permanent rivers or streams.

7.2 Schematic transect of sites in order of relative abundance of moisture for plant growth in Sudan (after J. Smith, 1949, Plate XI)

Physiognomic vegetation classifications are the third method. These are generally the easiest to quantify and relate to land classes. They are based not on floristics but on the physical form of plants, mainly their outward superficial appearance resulting from life form. Such schemes have wider application than the others because they are of practical value not only to users of the vegetation itself such as foresters, but also to non-specialists who view vegetation mainly as an obstacle, a source of cover, or an amenity. This method demands further consideration.

Classifications of vegetation physiognomy

Classifications of vegetation physiognomy have been approached from three points of view: from the ground, from the air, and from the aspect of practical applications.

The pioneering scheme for the ground description of plant physiognomy was by Kuchler (1949), but this was considerably developed by Dansereau (1958) whose structural classification has formed the basis of most recent work. Its essence is the definition of all plants in terms of six different parameters: life form, size, leaf function, leaf shape and size, leaf texture, and ground coverage. A simplified version is shown in Fig. 7.3.

CODE AND LEGEND FOR DEFINING VEGETATION PHYSIOGNOMY
(after Phillips, 1965, p. 33)

1. Life form
   T. ○ trees
   S. ♀ shrubs
   H. ▽ herbs and grasses
   M. ◠ mosses and other bryoids

2. Size
   1. more than 25 metres
   2. 10–25 m
   3. 8–10 m.
   4. 2–8 m.
   5. 0·5–2 m.
   6. 0·1–0·5 m.
   7. 0–0·1 m.

3. Function of leaves
   d. ▭ deciduous
   e. ▦ evergreen
   s. ▨ succulent
   l. ▥ leafless

4. Leaf shape and size
   n. ◠ needle
   g. ◊ grass
   m. ◇ medium, small (up to 20 cm²)
   b. △ broad, large
   v. ∿ compound

5. Leaf texture
   f. ▨ filmy
   z. ▭ papery
   x. ▭ hard and tough
   k. ▨ succulent

6. Coverage
   i. Discontinuous or interrupted:
      1. 0–20%
      2. 21–40%
      3. 41–60%
   c. Continuous:
      4. 61–80%
      5. 81–100%
   b. Barren or rare
   p. Tufts or groups

7.3 Code and legend for defining vegetation physiognomy (after E. Phillips, p. 33)

Sample areas representative of terrain units are chosen according to the size of the plant dominants involved. Squares with sides of 10 m are generally required for trees, 4 m for shrubs, 1 m for herbs and grasses and 10 cm for mosses and lichens. The plants within the sample area are counted and represented on graph paper with seven subdivisions of the vertical scale, to correspond to the subdivisions of plant height and with enough subdivisions of the horizontal scale to accommodate all the plants in the chosen sample area. A diagram is drawn by magnifying the life form and leaf shape and size symbols to accord with the sizes of the plants they represent to give a completed structural chart of the type shown in Fig. 7.4. A more refined and specialised form of this structural chart system has been worked out for the US Army by Addor (1963). This has a larger number of height classes and introduces additional sets of symbols for crown shape and diameter, stem diameter, habit, hardness, and succulence, branching type and height, root structure, height of emergence and spread, leaf size, shape, texture, and condition, armature (i.e. spines, stings, etc.), distribution patterns and such special elements as epiphytes and fallen trees.

A similar study for the Canadian army by Parry, Heginbottom, and Cowan (1968) considered the vegetation of Camp Petawawa, Ontario, from the point of view of its penetrability by military vehicles. This is related to stem diameter and stem spacing, of which only the latter can be readily assessed from aerial photographs.

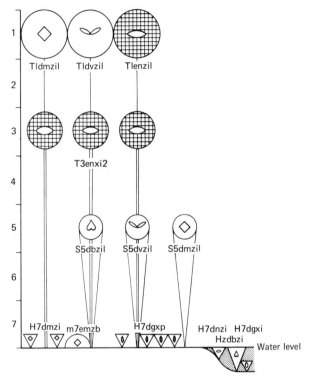

7.4 Example of a vegetation structural chart by the method of Dansereau

The increased use of aerial photography has focused attention on the appearance and classification of vegetation not only in profile view on the ground but also as it appears in plan view from the air. This is especially important from the military point of view because of the value of vegetation cover in concealment. In the Camp Petawawa study, this characteristic was found to be due to three aspects of the plant cover: height, canopy closure, and seasonal changes. Only the first two were evaluated but all three can be determined from the air.

Scale is the main determinant of the degree of detail with which it can be viewed, and each structural element has a critical scale below which it cannot be identified.

An example, which can be taken as typical, of the sort of semi-quantified physiognomic classification that is possible is that employed in the Aitape-Ambunti area of North-west New Guinea using 1:50,000 black and white aerial photography (Heyligers, 1968). Three types of grassland were recognised: mid-height (2 to 6 ft), tall (over 6 ft) and scrub, and also palm vegetation. The last was subdivided into low, mid-height, and tall, with limits at 50 ft and 100 ft. The upper canopy was classed as open, irregular, or closed, and the lower storey was classified in accordance with crown size, regularity of cover, and presence of a particular species (sago palms). This and other classifications from aerial photographs, such as that of Howard (1970b), are only partially quantitative, and cannot become more so except to the degree that plant species and structures, rather than merely photo tone and texture, can be recorded. Howard (1970a) has made a step in this direction with the design of a 'stereoprofiler' by which land and vegetation profiles can be derived automatically from aerial photographs.

Practical applications

Grabau and Rushing (1968) describe a practical development of this descriptive system for vegetation taxonomy for using quantitative manipulation by computers. It uses a physiognomic classification of the Dansereau type but arbitrarily determines sample area sizes in terms of the *structural cell* which, simply stated, is the minimum area showing all the internal spatial variation in an assemblage. In practice this means a circular site enclosing twenty plants of the size with which the survey is concerned. Only physiognomic parameters which are mutually independent are selected and these are recorded in such a way as to minimise transcription errors. Field observations are made on a form which can subsequently be punched out and itself used in the computer.

This computer-based approach has important implications for terrain study. It has been related practically to cross-country mobility problems by the US Army, who feed information on tree stem girths and spacings, representing the vegetational environmental controls, into a mathematical model for predicting cross-country vehicle speeds. The method is as follows: the computer reads numerical values approximating to the tree height, crown radius, and stem diameter from the punched sample cards and automatically calculates the stem sizes and spacings to be expected in the area from which the sample was drawn. The data can also be printed out in visual form as a cumulative graph relating stem diameter classes to number of plants in a given area.

This system has applications in forestry and ecology where information is required on the spatial geometry of plant assemblages. It could, for instance, be used to determine such facts as the degree of association of two species or the spatial arrangement of stem diameter classes. An easy way to achieve such objectives would be to print out annotated maps of the sample areas, which should each be equivalent to structural cells, using data on stem size, compass bearing, and distance of individual plants or plant groups from the centre of the sample area.

Physiognomic vegetation classifications of the type considered have also been recognized on aerial photographs and used for detailed mapping of selected areas chosen as representative of the terrain in Thailand across which military vehicles must operate (Broughton and Addor, 1968).

The most recent development in the field is the automation of the interpretation of vegetation from aerial photographs which can, for instance, improve the capacity and speed of forest resource mapping. One method is described by van Roessel (1971). A stereopair of panchromatic aerial photographs are scanned and digitised with a helical optical scanner. In their digitised form the photographs can be thought of as 1 728 × 1 728 arrays of square spots each associated with a grey value on a 0–7 scale. These are divided into cells of 20 × 20 spots each. Each cell can be interpreted for forest type by means of a statistical classification of the spot distributions, and the digitised information used to print out a map of forest types.

8
Basic geomorphological schemes

Suggested landscape units

Physiographic regionalisations have been undertaken by many writers since the end of the last century. Most of these have been geographers, geomorphologists, engineers interested in transport and communications, or land resource scientists. Reviews of their work have been made by Christian and Stewart (1964), Wright (1967), Beckett and Webster (1969), and Perrin and Mitchell (1970).

In the United States, the challenge of rapid expansion and newly developing land created an interest and stimulated activities in geographical studies and methods in the late nineteenth and early twentieth centuries. Joerg (1914) reviewed the various approaches to regional subdivision of the USA to that time. Bowman (1914) subdivided the country into *physiographic types* which he related to land use. He recognised the relationship between physical elements, human activity, and economic values, and that man's use of the land in each locality was controlled by the dominant recurrent physical elements, notably topographic configuration, water supply, and climate. Heath (1956) noted that Bowman's ideas influenced many surveys between the wars. In 1916 the Association of American Geographers established a committee under Fenneman to define the physiographic regions of the country. They used *section, order,* and *division* as their higher units. In 1933, Veatch gave these ideas a practical form when he classified agricultural lands in Michigan into *natural land types* based on soil topography, natural drainage, and native vegetation, which he foresightedly claimed would have as permanent a value as would soil or geological maps of the area.

The pioneering study in Britain was that by Bourne (1931) who defined a *site* as a unit which would for all practical purposes provide throughout its extent similar conditions of climate, physiography, geology, soils, and edaphic factors. Sites recurred in associations he called *regions*. He also pointed out the help which could be obtained from aerial photographs. Wooldridge (1932) described *flats* and *slopes* as the ultimate units of relief and defined the generic term *facet*. Unstead (1933) recognised a hierarchy of units by defining the terms *feature*, *stow*, and *tract* in order of increasing size for the lower order subdivisions of landscape. Milne (1935), working in East Africa, sought a soil sampling unit which would allow a complex entanglement to be mapped at small scale. He used the term *catena* for toposequences of soil which recurred on the same parent material and climate. Linton (1951) combined Unstead's with Fenneman's terms (replacing feature with site) into a hierarchy of morphological regions. Jenny (1958) subsequently suggested the term *tesserae* for units which seem to be somewhat smaller than facets.

Brink *et al.* (1965) coordinated these concepts into a hierarchy of units of ascending size: *element, sub-facet, facet, recurrent land pattern* (subsequently abandoned in favour of the Australian term *land system*), *land region, land province*, and *land division*, whose sizes and interrelationships are summarised in Table 8.1. The Australian work is considered in detail in chapter 9. It is sufficient to state here that the contribution has mainly come since 1946 from two departments of CSIRO: the Division of Land Research and Regional Survey and the Division of Soil Mechanics. The former introduced the terms *land unit* as well as *land system* (Christian and Stewart, 1964) and the latter used the words *pattern, unit*, and *component* for describing physiographic divisions of decreasing magnitude (Aitchison and Grant, 1967).

Canada, like Australia, has made extensive use of aerial photography for the evaluation of large undeveloped land areas since the war. Hills (1942, 1949, 1950) and Hills and Portelance (1960) mapped *land types* in parts of Ontario as a basis for planning future land developments, the underlying concept of which was defined by Lacate (1961). Significantly the criteria used in defining land types involved not only soils and bedrock but also the vegetation pattern including proportions, types, and distributions of the species within the forest ecosystem. At a wider scale, Hare (1959) described a reconnaissance survey of Labrador – Ungava in which sample areas were studied on aerial photographs to obtain a key of surface types, and then large areas were mapped using the key. An attempt was made to subdivide the types rather than merely to recognise and map them as complexes.

The Soviet approach to the problem derives from the pioneer work of Dokuchaiev, Beng, Ramensky, and others, and has been recently described by Solntsev (1962) and Vinogradov *et al.* (1962). The system appears to be remarkably similar to those from other parts of the world, notably the Australian. Definitions of land types are in terms of relief, soil, bedrock, microclimate, and habitat conditions such as moisture, salinity, and vegetation. Economic significance is also noted. The basic unit is the *facies*, which is the smallest elementary physical geographic subdivision of the landscape, having constant site and habitat conditions and a single 'biocoenosis', and which is therefore homogeneous genetically, physiognomically, and morphologically. Prokaiev (1962) opposed the too rigid definition of facies because of the variety of land uses

47

concerned with them and their inevitable internal variability. Facies are repeated in regular patterns in the landscape.

The next larger units in order of increasing size are *zveno, sub-urochishche, urochishche,* and *mestnost.* The urochishcha (plural of urochishche) are usually clearly reflected on aerial photographs and are defined as associations of facies with homogeneous substrate, common drainage, transfer of solid material and chemical elements. They can be simple or complex. They are regarded as complex if they have an association of facies related to a single relief form with sharply changing substratum, or an alternation of small relief forms, such as sand hillocks.

TABLE 8.1 Definitions of land units

NAME	DEFINED AS INCLUDING:	APPROX. RELEVANT MAP SCALE
Land zone	major climatic region	none given for land zones to be mapped on their own
Land division	gross form expressive of continental structure	1:15M
Land province	an assembly of surface forms and other surface features on a scale expressive of a second order structure or of a large lithological association	1:5M–1:15M
Land region	a small range of surface form and properties expressive of a lithological unit or a close lithological association having everywhere undergone comparable geomorphic evolution	1:1M–1:5M
Land systems	a recurrent pattern of genetically linked land facets	1:250 000–1:1M
Land facet	one or more land elements grouped for practical purposes; part of a landscape which is reasonably homogeneous and fairly distinct from surrounding terrain	1:10 000–1:80 000
Land element	simplest part of the landscape, for practical purposes uniform in lithology, form, soil, and vegetation	not to be mapped

In Germany, earlier concepts of *Fliese, Physiotop,* and *Landschaftszelle* have been more recently systematised into a hierarchical arrangement by the Institut für Landeskunde at Bad Godesberg (Schneider, 1966). The smallest basic unit is the *ecotope complex,* followed in ascending order by the *Fliesengefüge* or *microchore,* which is the main natural geographic unit, then the *naturräumliche Hauptenheit* or *mesochore*, which form the constituent parts of the larger *Gruppen naturräumlicher Hauptenheiten* or *macrochores*.

In Japan, a similar series of units identified by morphological development and genetic features, has been evolved. In order of increasing size these are *landform type, series, association, section,* and *province.* Implicit in all these systems is the concept that there is a small basic indivisible unit of terrain. Linton, in 1951, pointed out that the ultimate units of relief could be regarded as 'flats' and 'slopes', small units of surface that were visibly either one thing or the other, and were not susceptible to subdivision on the basis of form. He called them *sites,* and they can be considered as identical with the *elements* of CSIRO–MEXE (Brink *et al.,* 1965). Hills (1950) emphasised the importance of less subjective criteria in determining the limits of these smallest units, and Wright (1967) has contended that they should be defined more specifically by clearly visible properties with genetic significance, e.g. the natural superficial materials, gradients, local relief, the character of neighbouring drainage channels and depth of bedrock.

Similarly, despite differences in terminology and definition, there is a general parallelism of thought about the occurrence of distinguishable units of landscape which can be ranked into a hierarchy. There is also broad agreement that this hierarchy contains two levels of paramount importance approximating to the CSIRO–MEXE concepts of land system and facet respectively.

Theoretical landform associations

Landforms have been identified and classified collectively as well as individually. Certain authors have suggested theoretical schemes or models of groups of land units which occur regularly in association. Milne's term *catena,* which has attained wide currency, referred to recurring toposequences of soils on hillsides in East Africa.

Following the classical genetic approach initiated by Gilbert and developed by W. M. Davis and his followers, certain general schemes have been developed for classifying hillslopes. Notable are those of Penck (1927), King (1962), and more recently Dalrymple, Blong, and Conacher (1968).

Penck (as quoted in Thornbury, 1954) considered landscape within the context of the earth's crustal instability and viewed hillslopes as being convex, concave or linear, depending on the dominant process they were undergoing. If uplift was dominant and erosion unable to keep pace with it, slopes were convex; if erosion was dominant they were concave; while if the two processes were in balance they were linear. In a landscape under the second of these conditions (the *absteigende Entwicklung*) Penck recognized two parts of the hillslope, an upper relatively steep slope (*Bösche* or *Steilwand*) and a lower, relatively gentle slope (*Haldenhang*), which generally met one another at a steep angle. Meyerhoff (1940) called the former the *gravity slope* and the latter the *wash slope*.

King (1962), from South African experience, suggested a subdivision of hillslopes into four units: *crest, scarp, debris slope*, and *pediment* each associated with distinctive processes of mass movement and water flow as shown in Fig. 8.1.

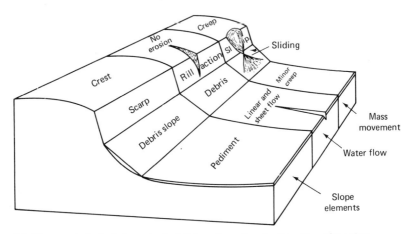

8.1 The morphological elements of a hillslope formed under the action of running water and of mass movement under gravity (Source: L. C. King, 1962, p. 137)

A more closely articulated scheme, based on data from New Zealand, has been suggested by Dalrymple, Blong and Conacher (1968). This divides the hillslope into a nine-unit land surface as shown in sketch form on Fig. 8.2.

Theoretical associations of a few landforms restricted to narrower distances and smaller ranges of lithological variety, and which are therefore at a lower degree of theoretical abstraction, are more frequent. Indeed, they are used in most geomorphological and soil surveys. A few illustrative examples may be quoted. Sand dunes in arid areas have been considered to be basically repetitions of only three surfaces: sand sheets, pack faces, and slip faces (Mitchell and Perrin, 1966). Coral reefs have been subdivided into types and constituent units on the basis of Australian examples by Fairbridge (1946–7) and volcanoes and volcanic landforms generally by Cotton (1944).

Comprehensive theoretical classifications of large areas

The earliest attempt to comprehend all land-forms over a wide area into a general classification seems to have been that by Passarge (1919). His scheme was hierarchical and remains one of the most comprehensive produced. It included five categoric levels: *type, class, order, family*, and *kind*. The two *types* represented landforms and coastal forms. The *classes* depended on the basic distinction between forms mainly suffering erosion and those mainly experiencing aggradation. *Orders* identified the main types of process: tectonic, volcanic, and eruptive. *Families* subdivided these on the basis of the main forms in which processes are expressed, e.g. the flexured, fractured, and faulted *families* of the tectonic *order*. Finally the *kinds* represented the degree to which the characteristics of the *family* were expressed, e.g. the distinction between symmetrical, asymmetrical, and overthrust *kinds* of the fault *family*.

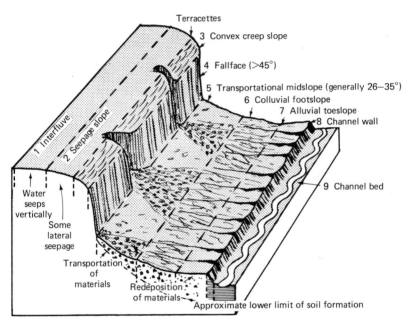

8.2 Hypothetical nine-unit land surface model (after J. Dalrymple, R. J. Blong and A. J. Conacher, 1968, p. 62)

A. D. Howard and Spock (1940) suggested a similar genetic system subdividing landforms into constructional and destructional types and then dividing these into diastrophic, volcanic, and depositional forms. The scheme of von Engeln (1942) was primarily structural, divided according to degree of consolidation and genetic origin of the resulting landform.

In order to fulfil the military requirement for intelligence about coasts, Putnam and others (1960) classified the natural coastal environments of the world. Again the fundamental distinction was into destructional and constructional landforms. The former were divided into plain, plateau, hill, and mountain types, each type being subdivided according to the dominant genetic process which formed them, emphasising especially the difference between stream-eroded and glaciated. The latter included alluvial, deltaic, ice-depositional, dune, coral, and volcanic types. Smaller shore features associated with the major coastal landforms were then considered. These were coastal dunes, tidal and flats, tidal woodlands, kelp beds, sea cliffs, marine beaches, and various types of reefs and barriers.

Also military in its inspiration was the classification of world deserts into physiographic units by Perrin and Mitchell (1970), which was carried out both at a land system and a facet scale. The main distinction was similar to that quoted above for coasts – between consolidated and unconsolidated rocks. Landforms on consolidated materials were subdivided into the lithological types which yielded different geomorphic forms. These were volcanic, crystalline, lineated metamorphic, consolidated siliceous sediments, consolidated calcareous sediments, laterite, evaporite, weakly consolidated coarse textured sediments, and weakly consolidated fine textured sediments. Each major constructional process by which new unconsolidated deposits are formed was considered definitive of land systems. These were: colluvial, alluvial, hydromorphic, coastal and eolian. An example of one of the thirty-nine *land system abstracts* with its constituent facets is given in Fig. 8.3.

Geometrical systems for small landforms

Geometrical classification schemes differ from those discussed in preceding sections in that they consider only the smallest land units, take no account of soil or lithology or any other terrain feature except surface geometry, and are essentially parametric rather than physiographic. They explore means of expressing surface form on maps and mathematically. Notable among them are the morphometric mapping of the 'Sheffield School' and the general mathematical analysis of microrelief which has been devised by Stone and Dugundji (1965).

The methods of the Sheffield School originated as a body of techniques evolved for slope measurement by Savigear (1952, 1956, 1962) and Young (1963), and developed into morphological mapping by Waters (1958), Savigear (1960), Bridges and Doornkamp (1963), and Curtis *et al.* (1965). These techniques have been most recently summarised and extended by Savigear (1965). They had two objectives: to determine the origins of small geometric subdivisions and discontinuities in the landscape and to train undergraduates to recognise landforms. The technique is used for mapping at scales of between 1:1250 and 1:63 360. All the following features are recognised:

Flat	Slope less than $2°$
Slope	Slope $2°–40°$
Cliff	Slope greater than $40°$

Facet	A plane surface in the landscape of any gradient
Segment	A curved surface in the landscape of any gradient, called concave if negative, convex if positive
Irregular facet or segment	Facet or segment with surface irregularities too small to map at field scale
Micro-facet	Facet too small to be separately mapped
Micro-segment	Segment too small to be separately mapped
Morphological unit	Facet, micro-facet, segment, or micro-segment
Break of slope	Discontinuity of ground surface
Inflection	Point, line, or zone of maximum slope between two adjacent concave or convex segments
True slope	Direction and amount of maximum surface slope of a facet or segment
Apparent slope	Direction and amount of slope measured in any other direction

Each of these features is marked on maps by a special symbol, some of which are shown on Fig. 8.4. Conventions are also supplied for representing boundaries which are themselves made up of complex micro-units. Young (1971) has since placed the definition of these units on a more quantitative and objective basis.

The mapping, classification, and quantification of micro-relief was investigated by Stone and Dugundji (1965) as part of the US Army Engineer Waterways Experiment Station program for the military evaluation of geographic areas, and for this reason their study was biased towards considerations of trafficability. Microrelief was somewhat arbitrarily defined as consisting of surface irregularities 3 inches to 10 feet high and 4 to 64 feet in lateral extent. Twenty-two micro-terrains in California were examined, ranging from a wavecut terrace to a boulder-free dry wash. These were mapped at the very large scale of 1:120 and a series of radial profiles were constructed through each area at intervals of 15°. These profiles were then evaluated to give parameters which could be regarded as indicative of surface roughness. The basic problem was that it was not possible to express roughness with a single parameter. It required four; each measured along every profile: average number of changes in level (M), average height (An) and steepness (P) of major relief features, and the extent to which there was periodicity (K). When all these were considered together as constituting the 'roughness vector', two further quantities could be determined: the avoidance factor rho, which indicated the difficulty experienced by a vehicle in crossing the terrain, and the cell length CL, which indicated the minimum distance along a traverse required to encounter all its normal variations.

The first four parameters were derived from a Fourier analysis of the profile curves within a 64 foot envelope or 'linear packet'. The most

P1

Defined by	CM
Date	1 10.64
Ref.	

PATTERN CARD NAME   Fresh lava cones and flows

CLIMATIC REGIONS   Hot Arid (Meigs E and A 12-34 classes)

RELIEF   Up to several hundred metres

VEGETATION   (General)

COUNTRY ROCK   Newer volcanic material of all kinds*

SURFICIAL DEPOSITS

PHYSIOGRAPHY
TECTONICS } Fresh lava flows
SOIL.

Altitude (M)

Continuation Card

LOCAL FORMS	Fr. Somaliland	Aïr Massif, Niger	Round Mt. Pisgah Amboy, Calif., U.S.A.	Black Mts. Ariz.	Aquarius and S. Artillery Mts., Arizona	Shuqra Area, Aden
No. AND NAME	(A)	(B)	(C)	(D)	(E)	(F)
COUNTRY ROCK				Eruptives over xline core.	Andesite and rhyolite.	
GEOREF 15° 1°		NH (16-19°30' N) HB-KE(7°30' - 9°30' E)		EJ (35°15'-36°15' N) FF-FG (114°20'-114°40'W)	EJ GE	

DIAGRAM (NORM AND RANGE)   BY: REF:

COGNATE FACETS AND CLUMPS

No.	NAME
F127	Fresh volcanic hillslopes.
C 4	Bossed Volcanics.
F 8	Fresh volcanic flows .

DISTINGUISHING FEATURES

Areas of fresh uneroded lava flow including recent cones

* Rocks generally include andesite, basalt, lava, diabase, phonolite, tachylite, trachyte, rhyolite, dacite, trap.

ASSOCIATED FEATURES   Surface covered with basaltic or similar boulders.

GENESIS (GENERAL)
AND TO INDICATE SENSE OF VARIATIONS   Fresh lava outpourings.

INTER-RELATIONS
COGNATE OR GRADED BOUNDARIES
(i.e. PATTERNS WHICH MUST OCCUR AS BOUNDARIES)
AND
NON-COGNATE BOUNDARIES
(e.g. THE SORT OF PATTERNS WHICH OCCUR AS BOUNDARIES
THOUGH NOT GENETICALLY LINKED.)

Generally ends abruptly over plains areas which it overlies. Its edge is usually marked by a lobate tongue several feet high.

COMMENTS AND GENERAL REFERENCES

Rodd, F. R. A Journey in the Air. Geographical Journal LXII, August 1923, pp. 81-102.
Birot, P. and Dresch, J. Une Coupe à travers le Hoggar Central. Bull. Assoc. Geog. franc. (Paris) 253-4 (1955), pp.158-60.
Clements, T. et al. 1951. A study of Desert Surface Conditions. Env. Prot. Res. Div. U.S. Army H.Q. Q.M. Res. and Dev. Comd. Tech. Rept. EP-53-Natick, Mass.
Parker, R. B., 1963. Recent volcanism at Amboy Crater, San Bernardino County, California. California Division of Mines and Geol. Spec. Rep. 76.

8.3 The subdivisions of one characteristic type of arid landscape: fresh lava cones and flows

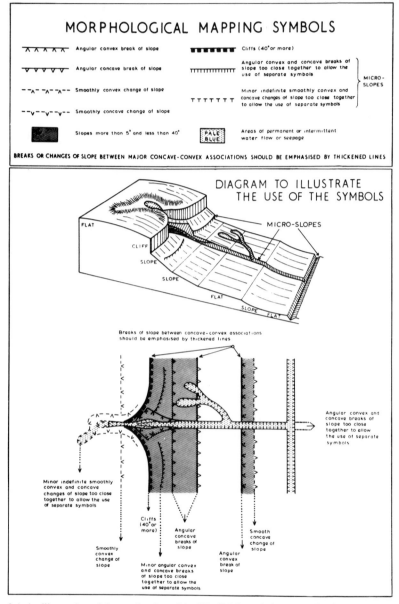

8.4 An illustration of the application of the 'Sheffield' symbols to some landscape patterns typical of a part of the Southern Pennines, England (Source: R. A. G. Savigear, 1965, p. 518)

generally useful of the vector components was the avoidance term rho, which was the product of P and M. Its magnitude was, therefore, an index of surface roughness, and inversely related to the traversability of the terrain in question.

Professional geomorphological mapping schemes

The first geomorphic mapping system was that suggested by Passarge (1919) and it remained the only example until the idea was developed in a number of countries, notably Poland, France, and Russia after 1950, followed by Czechoslovakia, Hungary, Belgium, Canada, and the Netherlands in the 1960s, whose contributions are reviewed by Saint Onge (1968). These differed considerably, and in 1960 a Subcommission on Geomorphological Mapping was set up by the International Geographical Union to standardise legends and ensure that there was an agreed scheme for showing the appearance and lithology (*morphology*), the dimensions and slope values (*morphometry*), the origin (*morphogeny*), and the age (*morphochronology*) of each form. The size of the problem can be seen from the fact that the Russian scheme, which gave attractive maps and was the most comprehensive, had over 500 items on the legend, but yet lacked any slope data.

Verstappen and Van Zuidam (1968) produced a flexible adaptation of previous schemes which provided codes for general purpose 'standard' geomorphological maps. These had a legend of 469 symbols divided among twelve groups representing the broad lithogenetic types of landscape: – volcanic, fluvial etc. They also provided codes for special purpose 'morpho-conservation' maps for engineering and conservation purposes, and 'hydro-morphological' maps for hydrological purposes. Fig. 8.5 reproduces a few of the subclasses and their symbols used in the latter. Generalised small scale land surface form maps have also been produced of North and South America and of the USA by Hammond (1962, 1964) and of the world by Murphy (1968). These are shown on Figs 8.6–8.9.

14. HYDRO-MORPHOLOGICAL MAPS
a. Surface water-natural

14·1	hydromophological units		14·11	vanishing riverbed	
			14·12	dry valley	
14·2	base flow bed		14·13	abandoned riverbed	
14·3	minor bed		14·14	capture	
14·4	bankful bed		14·15	bank storage	
14·5	flood limits		14·16	waterfall with height	8 m
14·6	stream order		14·17	rapids	
	divide lines		14·18	rapids in gorge	
14·7	major			ponor	
14·8	minor		14·19	perennial	
14·9	flood overflow		14·20	seasonal	
14·10	current direction		14·21	fossil	

8.5 A sample page from the ITC legend for geomorphological mapping (Source: H. Th. Verstappen and R. A. Van Zuidam, 1968, p. 44)

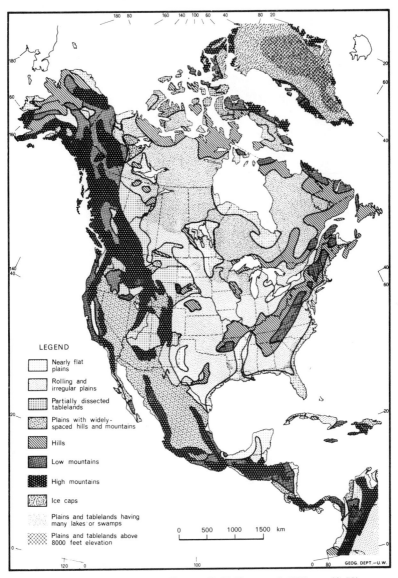

LEGEND

- Nearly flat plains
- Rolling and irregular plains
- Partially dissected tablelands
- Plains with widely-spaced hills and mountains
- Hills
- Low mountains
- High mountains
- Ice caps
- Plains and tablelands having many lakes or swamps
- Plains and tablelands above 8000 feet elevation

0   500   1000   1500 km

GEOG. DEPT.—U.W.

8.6 Terrain types of North America (Source: E. H. Hammond, 1962, pp. 69–75)

59

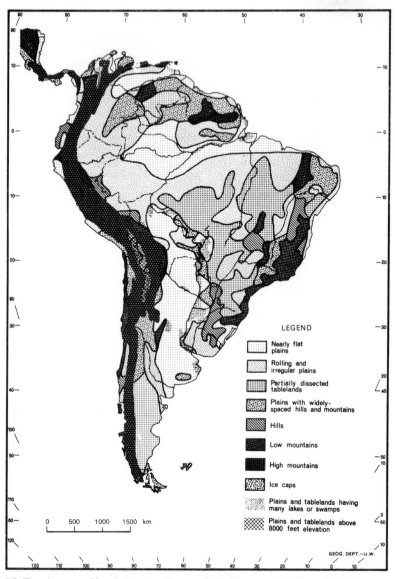

LEGEND

Nearly flat plains

Rolling and irregular plains

Partially dissected tablelands

Plains with widely-spaced hills and mountains

Hills

Low mountains

High mountains

Ice caps

Plains and tablelands having many lakes or swamps

Plains and tablelands above 8000 feet elevation

GEOG. DEPT.—U.W.

8.7 Terrain types of South America (Source: E. H. Hammond, 1962, pp. 69–75)

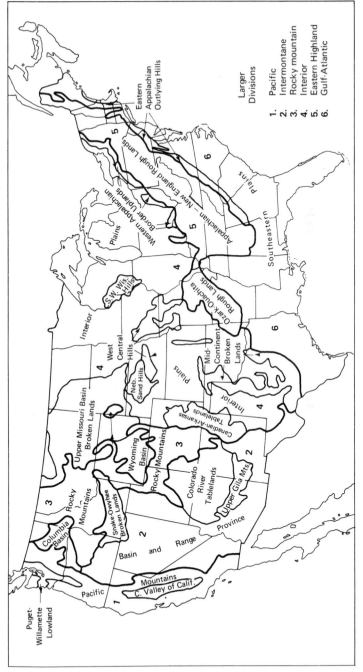

8.8 Summary of the principal subdivisions of land-surface form in the forty-eight states, USA (Source: E. H. Hammond, 1964, map supplement no. 4)

LEGEND:
The 1st letter indicates geological age: A Alpine, C Caledonian and Hercynian,
    G Gondwana, L Laurasia, R Rifted Shield areas, S Sedimentary cover outside
    Shield exposures, V Volcanic.
The 2nd letter indicates landform type: M mountains, W widely spaced mountains,
    T high tablelands, H hill and low tablelands, D depressions or basins, P. plains.
The 3rd letter indicates climate:
   i. Ice caps at present. ●●●●●
   w. Wisconsin or Wurm glaciated areas. ●—●—●
   g. Pre-Wisconsin, pre-Wurm and undifferentiated Pleistocene
      glaciated areas. ― ― ― ―
   h. Humid landform areas.
   d. Dry or arid landform areas.
Division between humid and dry landform areas. ⌒
Major oceanic rift and fault lines. ――/――
Continental shelf. ×××××

8.9 Landform map of the world (somewhat simplified from R. E. Murphy, 1968)

Part Two

Practical Systems of Terrain Evaluation

9
Systems used in soil science, agriculture, pasture and forestry

The subdivision of terrain evaluation schemes in the following chapters is based on the main types of professional land user. This subdivision is arbitrary, especially in view of the aim expressed earlier in the book of suggesting an integrated scheme of general value to all interests. It is justified on the ground that although the techniques of integrated survey have developed from the convergence of a number of applied sciences, these remain today far from integrated in fact. Research and professional practice are still separated into the main disciplines of agriculture, engineering, planning etc., and much remains to be done in bridging the gaps between them. The time has not yet arrived at which it is possible to treat the conceptual development in an integrated manner, and there is no adequate alternative to an analysis which subdivides the methods according to professional interests.

Soil science

Soil classification and mapping has had a history differing fundamentally from that of terrain, but the increasing number of practical disciplines which must include consideration of both, and the revolutionary advances in the techniques of remote sensing of application to both, have combined to make them convergent.

Early schemes of soil classification were artificial and parametric in that they selected soil characteristics indicative of soil fertility and mapped these. A natural classification only became possible when it was recognised that soils were natural entities with distinctive morphologies. The first such classification was that of Dokuchaiev,[1] who recognised them as independent, subaerial, natural bodies, with properties reflecting the effects of local and zonal soil-forming agents. Sibirtsev, his closest collaborator, introduced the concepts of zonal, intrazonal, and azonal soils to embrace soils which reflected respectively the normal climate of an area, the influence of a dominating local factor (usually excess water), and an absence of normal pedological development.

1. For historical summary of the development of soil classification, see Soil Survey Staff (1960).

These concepts were soon adopted elsewhere and form the basis of almost all soil classifications today, especially where they have been most fully developed in Europe, North America, and Australia. The most significant recent schemes of soil classification which aim at worldwide comprehensiveness are those of FAO (United Nations 1971+) and the US Department of Agriculture 7th Approximation (Soil Survey Staff, 1960) and its subsequent modifications. These attempt to impose a more rigid framework onto previously somewhat vaguely defined soil classes by giving them more exact and often quantitative parametric definitions. Parametric classifications are still in use today. Detailed soil surveys of small areas of land suffering from defined land use hazards are usually of this type. Early soil maps of the Sudan Gezira, for example, were based on an index called the 'sodium value' which was closely correlated with cotton yields (Tothill, 1952). Numerical classes of soil texture, salinity, or alkalinity, have been used extensively in *ad hoc* surveys for irrigation in developing countries, e.g. by Hunting Technical Services in Iraq, Pakistan, Kenya, and elsewhere (1954+).

Physiographic methods

Soil series, which are the basic units of soil mapping, are more uniform and represent a finer subdivision of the landscape than do the terrain units of which they form a part. Only where land is little developed or of relatively low economic value do the latter form an adequate classification of landscape for agricultural purposes.

Outstanding examples of their use are the approaches employed by Australian CSIRO Division of Land Research and Regional Survey and the British Directorate of Overseas Surveys Land Resources Division. These are hereinafter abbreviated to LRRS and DOS respectively.

Although the work of LRRS is here considered under agriculture, pasture, and forestry, it has always viewed terrain from the point of view of the whole range of land use interests, including also the engineering aspects of soils, water supply, minerals, wildlife, fisheries, harbours, scenery, tourist and recreational attractions, etc. But although this range of concern is wide and varied, it is essentially confined to natural resource factors which have relatively fixed geographical location and extent and are amenable to geographical forms of analysis. They are thus clearly distinct from other economic resource factors, such as capital and labour.

The background and methodology of integrated surveys carried out by LRRS (and other bodies) have been summarised by Christian and Stewart (1964). Extensive resource surveys in undeveloped parts of Australia, Papua, and New Guinea were begun in 1946, using the aerial photography and techniques developed during World War II. These surveys were based on *land systems* and *land units* which aimed at being both basic and functional divisions of landscape. Land systems were

(14) BALBI LAND SYSTEM (30 SQ MILES)

Active or recently active volcanoes.

Geomorphology.—Very recent volcanic land forms including lava flows, debris slopes, and scarps forming the margins of explosion craters or of spines. Mt. Bagana is an active craterless lava cone; Mt. Balbi is a group of cratered active or inactive ash or scoria cones and spines; Lake Billy Mitchell is a large explosion crater; an unnamed peak east of Mt. Bagana is a dissected lava-flow cone. Lakes and ponds occur both in craters and in valleys blocked by lava flows.

Terrain Parameters.—Altitude: H.I., V; min., 1000 ft; max., 8500 ft. Relief: very high (1500 ft). Characteristic slope: very steep. Grain: very coarse (5000 ft). Plan-profile: 4.

Geology.—Andesite, hornblende–andesite, and hornblende-bearing basalt, as lava and ash; Recent.

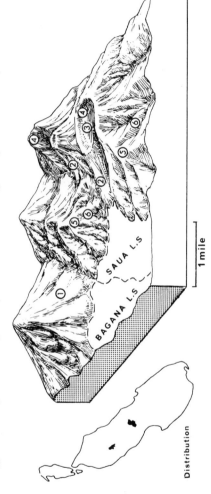

1 mile

Distribution

Land Unit	Area (sq miles)	Land Form	Soil	Land Class; AASHO Soil	Vegetation
1	8	Lava flow: sinuous stream of rock up to 2 miles long and 1000 ft wide expanding distally to 3000 ft; steep axial slope (17–30°), increasing at the terminus to very steep (42°); cross-section convex except in upper parts which are irregularly concave; commonly bounded by narrow, very steep, marginal ridges up to 50 ft high	Blocks	VIIIst$_8$ B	Bare except for terminal slope with mixed herbaceous vegetation (Lycopodium–Gleichenia)
2	3	Debris slope: long (2000 ft); concave; mainly gentle or moderate (2–20°); traversed by stream beds up to 50 ft wide, locally incised 100 ft	Mainly stones and boulders / Locally ash soils: grey fine sands (stony phase) / or / alluvial soils: shallow grey mottled sands	VIIIst$_8$ B+C Vle$_3$,st$_6$,n$_2$ A3+C IVd$_3$,f,n$_2$ A3	Bare; mixed herbaceous vegetation (Lycopodium–Gleichenia, mountain herbaceous vegetation); savannah; and scrub (Cyathea–Bambusa, mountain scrub)
3	½	Crater floor: nearly level; up to 500 ft diameter; 1 ft channelled microrelief; ponds	Ash soils: grey fine sands	VIId$_7$,n$_2$ A3	Bare or with mountain herbaceous vegetation; probably some bogs
4	2½	Scarp: medium length to long (500–2000 ft); irregular; precipitous; commonly with waterfalls	Rock outcrop or ash soils: grey fine sands (stony phase)	VIIIr$_8$ R$_H$ VIIIst$_8$,n$_2$ A3+C	Bare or with mountain herbaceous vegetation
5	1	Ridge crest: knife-edged or very narrow; very uneven; steep crestal slope	Ash soils: grey fine sands (stony phase)	VIIIst$_8$,n$_2$ A3+C	As unit 2, locally palm and pandan vegetation (Gulubia–Pandanus)
6	15	Erosional hill slope: short or medium length; straight; very steep to precipitous	Boulders or ash soils: grey fine sands (stony phase)	VIIIst$_8$,n$_2$ B or A3+C	As unit 2, locally mountain low forest

Population and Land Use.—Nil.
Forest Potential.—No forest. Access category III.
Observations.—2, plus 4 aerial observations.

9.1 An example of the CSIRO Division of Land Resources and Regional Survey method of defining and describing terrain: the Balbi Land System covering 75 sq km (30 sq miles) on Bougainville Island (Source: CSIRO, 1967, p. 34)

recurrent patterns of landforms, soils, and vegetation recognisable on aerial photographs. Land units were their constituent subdivisions. Land system surveys were presented in the form of maps at scales of 1:250 000 and 1:1M accompanied by block diagrams showing the interrelations of the land units and tabulated summaries of their form, soil, and vegetation properties. Fig. 9.1 gives an example of one land system from the Report on Bougainville (CSIRO, 1967). Shortage of staff, funds, and information made it impracticable to map land units at their relevant scale of about 1:50 000–1:100 000, and so although they are the main vehicle for data storage, land systems are the basic mapping unit. More than twenty such reports had been published by the end of 1967 covering in total about 650 000 square miles in Australia and 30 000 in Papua, New Guinea, and Bougainville, a substantial proportion of the total area of these countries, as can be seen from Fig. 9.2.

DOS use essentially the same method for planning agricultural development in extensive unsurveyed regions. Land system surveys for reconnaissance are carried out at scales of 1:250 000 or 1:500 000, intensive land resource assessments at 1:50 000 or occasionally 1:25 000, and development studies at 1:10 000 or larger. These can be done on an 'integrated' basis, i.e. including all resource characteristics or on 'single aspect' basis where only one is studied. Reconnaissances have been carried out in recent years on the former basis in Bechuanaland, Nigeria, Lesotho, Malawi, and British Solomon Islands, and on the latter in Malawi and Botswana (Ministry of Overseas Development, undated). Intensive resource assessments can likewise either be conducted on an integrated or a single aspect basis. Studies in Malawi and Fiji are examples of the first and others in New Hebrides, Fiji and Kenya, of the second type. Detailed development studies generally relate to the cultivation of arable crops on relatively good soils or the 'working plan' for detailed forest exploitation. There are examples of such projects in Tanzania and Christmas Island.

Other individuals and organizations have used the land system method to guide rural development. Taylor (1959) has applied it in Nicaragua, retaining the term 'land unit' but using 'land type' instead of 'land system'. In Pakistan, Thirlaway (1959) used it in the Isplingi Valley and Wright (1964) on a Unesco project in the Nagarparkar area. Photographic Survey Corporation (1956) classified all the Indus plains on a landform basis. Parts of India have been covered by the Basic Resources Division of the Central Arid Zone Research Institute at Jodhpur, India, and Hunting Technical Services (1954+ and Fig. 7.1) have employed the landscape approach for range classification in Jordan and on a regional development survey in Jebel Marra, Sudan (*ibid*).

A practical appraisal of the value of the method for regional planning was made by Renwick (1968). The area chosen was the Hunter Valley,

New South Wales. This area was useful because it was a well-defined river basin which was also subject to study by a special Research Foundation, established in 1956, to assess the investment structure and potential of the region. The area was regarded as an illustrative sample of the whole path of economic growth in Australia. Coordinated studies were made of land, water, and economic structure. The area was divided for study purposes into rural and urban parts. The rural study aimed to determine how to reallocate the agricultural resources to optimise growth. The area was divided into land systems, and vegetation cover and land utilisation data were related to these. Water resources were studied in terms of subcatchments, which did not necessarily

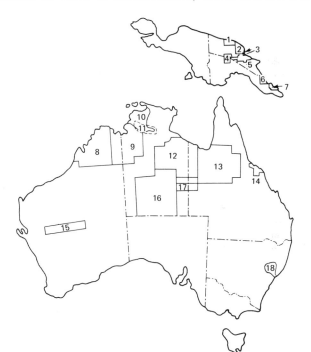

1. Wewak–Lower Sepik (4500 sq. miles)
2. Lower Ramu–Atitau (4500 sq. miles)
3. Cocol–Upper Ramu (3500 sq. miles)
4. Wabag–Mendi (6000 sq. miles)
5. Goroka–Mt Hagen (4000 sq. miles)
6. Buna Kokoda (2500 sq. miles)
7. Wanigela–Cape Vogel (2000 sq. miles)
8. Kimberley (81 000 sq. miles)
9. Ord–Victoria (80 000 sq. miles)
10. Katherine–Darwin (27 000 sq. miles)
11. Tipperary (73000 sq. miles)
12. Barkly (120 000 sq. miles)
13. Leichhardt–Gilbert (120 000 sq. miles)
14. Townsville–Bowen (6000 sq. miles)
15. Wiluna–Meekatharra (25 000 sq. miles)
16. Alice Springs (144 000 sq. miles)
17. Georgina Poison (24 000 sq. miles)
18. Hunter Valley (8500 sq. miles)

9.2 Areas in Australia and New Guinea surveyed by Division of Land Research and Regional Survey to end of 1967 (Source: J. A. Mabbutt and G. H. Stewart, 1967, p. 97)

coincide with land systems. Urban land utilization was studied separately. The importance of the study lay in the proof it gave of the value of the landscape system in a small highly settled area where rural land use and agricultural productivity were the keys to development.

Parametric systems for agriculture, pasture, and foresty

There are so many schemes for parametric classification of land for these purposes that all that is possible is to outline some of the more familiar types. They can be divided into those concerned with existing land uses and those concerned with land potentials.

Land use mapping is carried out in almost all countries. A pioneer was the Land Utilization Survey of Great Britain produced in the years following 1930 under the inspiration and direction of L. D. Stamp who, .vith Willatts, summarised the methods and first results in 1935. The whole of Great Britain was covered by mapping at 1:63 360 with a report for each county. The six classes used were simple and empirical: forest and woodland; meadowland and permanent pasture; arable or tilled land; heathland, moorland, commons, and rough hill pasture; gardens, allotments, orchards, and nurseries; and land agriculturally unproductive. An example of a part of a map is given in Fig. 9.3.

The second Land Utilization Survey of Great Britain was commenced in 1960 under the direction of Alice Coleman (Coleman and Maggs, 1965). It is being produced at 1:25 000 and so requires 843 sheets to cover England and Wales alone, of which only 109 are complete at the time of writing (May 1971). The basic classification adopted is essentially the same as in the first survey but each class has been divided into a number of subclasses so that the map can be interpreted at two levels of generalisation. For instance, arable land is subdivided into cereals, roots, and green fodder crops, and woodland into coniferous and deciduous types. A small part of a map is reproduced in Fig. 9.4.

Areas in Africa have been covered with land use surveys by DOS e.g. that of the Gambia at 1:25 000 (1958), and also by Hunting Technical Services (1954+) and other organisations. Parts of Canada have also been mapped on a similar basis (Canada, Department of Mines and Technical Surveys, 1959).

Appraisals of the agricultural value and potential of land are usually carried out by organisations concerned with soil surveys, a number of whom have brought out general schemes for classifying the economic value and potential of the soils considered. These schemes are all parametric and are generally based on 'penalty points' systems by which land values are determined by the number and degree of limitations affecting land use.

The first such scheme to appear was that of the USDA Soil Conservation Service (Klingebiel and Montgomery, 1961), which is widely used today. To quote Bibby and Mackney (1969) 'it assesses land

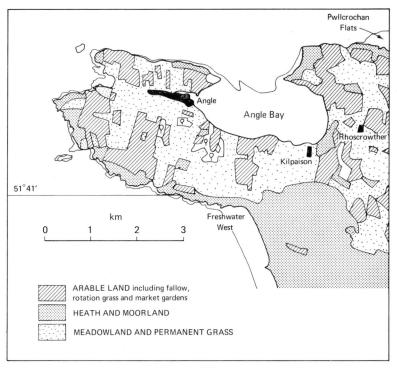

9.3 First Land Utilisation Survey mapping. Part of Pembroke and Tenby Sheet

capability from known relationships between the growth and management of crops and physical factors of soil, site, and climate'. Land suited to cultivation and other uses is included in classes 1–4 and land not generally suitable for cultivation and of only limited use for other purposes in classes 5–8. Class 1 land has a wide range of uses with few limitations while the remaining 7 classes suffer from increasingly severe limitations and are progressively less flexible. Capability subclasses are defined on the physical factor or factors limiting production, each of which is indicated by the use of a letter subscript attached to the relevant class number. These identify the capability units which are the lowest category of the classification, and group the soils capable of growing the same kinds of crops and requiring the same management. Long-term estimates of crop yields for individual soils are qualified by the statement that they should not vary from the true values by more than about 25 per cent. Table 9.1 summarises the criteria for the main classes.

The subclasses are categorised as being due to *erosion*, where susceptibility to this is the dominant hazard; *excess water*, where drainage is poor, the water table high, or where there is an overflow danger; *soil limitations*, where the rooting zone is restricted by such factors as

71

shallowness, stones, low moisture holding capacity, low fertility or high salinity or sodium, and *climatic limitations* where the climate (temperature or drought) is the only major hazard to use.

The same scheme has been adopted both in the Canada Land Inventory (1965) and by the Soil Survey of England and Wales (Bibby and Mackney, 1969). Both these countries have eliminated the fifth class, which allows mainly for wet soils in level sites poorly adapted for arable crops. In Britain, a further subclass has been added for *gradient and soil pattern* limitations on land use.

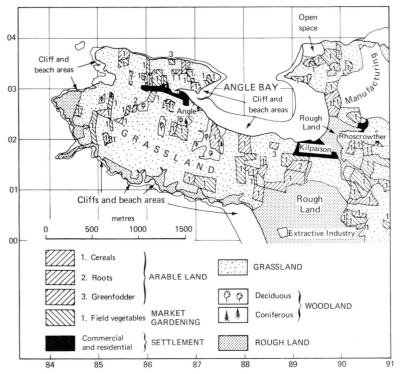

9.4 Second Land Utilisation Survey mapping. Part of Milford Haven Sheet

The Agricultural Land Classification of England and Wales, published by the Ministry of Agriculture, Fisheries, and Food (1968), adopts a simple scheme which uses only five grades to classify land for its agricultural potential, ranging from Grade 1 with very minor or no physical limitations, to Grade 5 with very severe soil, relief, or climatic limitations which is now under grass or rough grazing. Maps are produced at 1:63 360 and 34 of the 117 required to cover England and Wales have been published between their first appearance in 1965 and the time of writing (May 1971).

TABLE 9.1. Summary of criteria for USDA land classes (after Klingebiel and Montgomery, 1961)

TYPE OF LAND USE	LAND CLASSES							
	I	II	III	IV	V	VI	VII	VIII
	(suited to cultivation and other uses)				(generally not suited to cultivation)			
1 Crops	+	+	+	+				
2 Pasture	+	+	+	+	+			
3 Range	+	+	+	+	+	+		
4 Woodland	+	+	+	+	+	+	+	
5 Wild life	+	+	+	+	+	+	+	+
LAND ATTRIBUTES								
6 Slopes	Level	Gentle	Moderately steep	Steep	Gentle	Steep	Very steep	Very steep
7 Erosion hazard	None	Moderate	High	Severe	Severe	Severe	Severe	Severe
8 Overflow danger	None	Occasional	Frequent	Frequent	Frequent	—	—	—
9 Soil depth	Ideal	Less than ideal	Shallow	Shallow	Shallow	Shallow	Shallow	Shallow
10 Soil structure and workability	Good	Somewhat unfavourable	—	—	—	—	—	—
11 Drainage	Good	Correctable by drainage	Very slow	Waterlogging	—	Waterlogging	Waterlogging	Waterlogging
12 Water holding capacity	Good	Moderate	Low	Low	Low	Low	Low	Low
13 Salinity	None	Slight to moderate	Moderate	Severe	—	Severe	Severe	Severe
14 Nutrient status	Good	Moderate	Low	—	—	—	—	—
15 Climate	Favourable	Slight limitation	Moderate	Moderately adverse	Unfavourable	Unfavourable	Unfavourable	—
16 Management practises required	Ordinary	Careful	Special	Occasional cultivation only possible	Cultivation not possible			
17 Stoniness	—	—	—	—	Some	Severe	Severe	Severe

In the USSR land evaluation is carried out for regional planning which includes consideration of industrial, civic, and urban uses in addition to agriculture and forestry (Ignatyev, 1968). There are two types of classification: basic and applied. The basic survey assesses the main natural properties of the land while the applied arranges these into functional groups to form land use units.

In central Australia, an evaluation of the grazing capacity of certain arid lands to assess rentals and avoid abuses in soil conservation and land management has been carried out in more detail (Condon, 1968). A rating was devised for each of the factors that influence grazing capacity. Aspects that increased it were given a bonus, those that decreased it, a penalty rating. The method was to select arbitrarily a base value from a widespread intermediate standard understood in New South Wales. This base was an area with eight dry sheep per 100 acres under an annual rainfall of 10 inches (254 mm). Ratings were then based on divergences from this standard in inherent fertility, moisture relationships, erodibility of soils, topography, tree density, drought forage availability, pasture type, health or condition of range, rainfall, and the occurrence of barren areas.

10
Systems for military purposes

R. Webster
P.H.T. Beckett

Military interest in terrain is of long standing, but its width and ramifications have increased rapidly of recent years as a result of the galloping sophistication of modern weapons.

Historically, the main concerns have been the situation, traversability and 'diggability' of terrain. Topography made it possible for 300 Spartans to hold the Persian army at bay in the narrow defile of Thermopylae, and the Black Prince gained success at Poitiers by pushing forward his light troops on to ground too soft for the French armour (Beckett, 1962). With the advent of gunpowder, questions of cover and intervisibility of sites became more important. The development of military interest in terrain since 1914 has resulted chiefly from the greatly increased size, range, and destructiveness of bombs, missiles, and projectiles, and the development of the internal combustion engine, tyred and tracked vehicles, and aircraft.

Military assessments of terrain have two aspects: strategic and tactical. At a strategic scale the concern is with the gross spatial distribution of economically important lowlands with their cities and associated transport lines, and also with the mountains, sea, and river barriers that divide them. Tactical assessments of terrain consider the landscape in more detail. They focus on four types of problem: those concerning visibility, cross-country mobility, the reaction of terrain to deformation, and the availability of water and constructional materials. It is important for military purposes to know the distribution of potential vantage points and areas of concealment which enjoy maximum visibility over an enemy, while suffering from minimum exposure to his oversight. Cross-country mobility or 'going' involve those aspects of terrain which determine its suitability for the off-road movement of vehicles, aircraft, and bodies of troops and for the reception of parachute drops. It is a function of the slope, evenness, hardness, and slipperiness of the ground surface and the number and nature of obstacles it contains. Thirdly, terrain must be judged in terms of its reaction to deliberate deformation such as the introduction of tent pegs and the excavation of trenches, caves, and dugouts which are free from the danger of caving and resistant to the penetration of projectiles. Finally, it is important to know the location and nature of sources of construction materials for buildings, roads, and airfields, such as gravel, sand and wood, and the general availability of potable water in usable quantities without the frequently attendant evil of a high water table which floods entrenchments and makes the surface soft and impassable.

Many of these factors can be illustrated from a brief consideration of one important example: the place of the Flanders lowland in the campaigns of both world wars. As Johnson (1921) has pointed out, this lowland was strategically vital not only in being the narrowest point on the great north European plain but specifically in covering the gap between the Artois and Ardennes barriers which is the only route into France from Germany without formidable topographic obstacles. In addition, it had strategic importance in its dense population, highly developed both agriculturally and industrially.

The plain itself has a most distinctive character. It is one of the lowest and flattest tracts in Europe. The clay of which it is composed is especially fine grained and impermeable, and usually wet. When in this condition, all movements are slowed and troops are exposed for a longer time to hostile observation. Shells lose effect; equipment and weapons become clogged, and the wounded suffocate. Excavations are difficult. Trenches will not stand up, soon fill with water, and become unusable unless shored with timber and pumped out. Water, though abundant, is contaminated and natural building materials are absent. All these factors adversely affect morale. There are numerous instances of the obstacle posed by the mud of Flanders to military activities in 1914–18, and it was an important factor in delaying the German advance and in permitting the British evacuation from Dunkirk in 1940.

Closer examination of the actual relief reveals other military implications of the terrain. Reference to Fig. 10.1 shows that within the clay

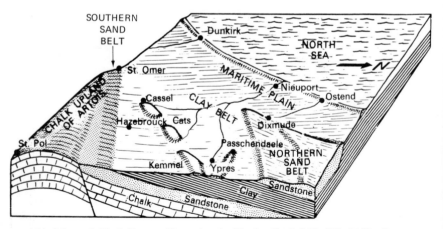

10.1 Schematic block diagram illustrating the Flanders Battlefield of World War I (after D. W. Johnson, 1921)

belt is an intercalated sandy bed which forms a cuesta with a south-facing escarpment. This is the Mont Kassel – Mont des Cats – Mont Kemmel upland, whose eroded backslope remnant is the Passchendaele – Messines ridge. Mont Kemmel, though barely 150 metres (500 ft) high, has been the key position in Flanders since Roman times. From its summit a wide view may be obtained stretching to the sea in the west, the Artois upland to the south, and almost to Brussels in the east. In World War I no less than six major battles were fought for its control and hundreds of thousands of British and German lives were sacrificed in attempts to gain it. At no time did either side possess the whole. Each attacked the other, not only on the surface, but also by tunnelling along the better drained intercalated sand lenses into the ridge and exploding the trenches of the other. The British used this technique to better effect than the Germans because they realised that water table levels continued to fall even after the commencement of the wet season.

Even in this brief summary, we see exhibited in Flanders almost all the military implications of a single type of terrain: the strategic location and dominating position of uplands and the rear cover provided by their possession, and the mobility, excavation, raw material, water supply, and drainage problems of clay lowlands.

Physiographic systems of military terrain evaluation

In Britain, as in other countries, notably Germany, France, Russia, and the USA, military interest in terrain has led to the development of methods for predicting it. Simple 'tank maps' to show the passability of ground were produced by the French army Service Géographique in 1918. Some 'going maps' were produced in World War II for special areas such as the Anzio beachhead in Italy, but it is only since that time that 'going' or cross-country mobility maps have been produced for larger continuous areas, for example, of Trucial Oman and Muscat (Directorate of Survey, 1959). These maps show the main subdivisions of landscape, patterns of settlement and land use at a general scale of 1:250 000, and are based on aerial photographic interpretation supplemented by data from published and unpublished sources. Like conventional soil maps, they are purely descriptive of the locality they represent and do not aim at any wider capability for extrapolation to analogous areas elsewhere.

It was MEXE (Military Engineering Experimental Establishment at Christchurch, Hants)[1] and the Soils Laboratory at Oxford who first conceived a system for predicting terrain information by storing it according to generic, uniform, operationally defined[2] physiographic

1. In 1971 the name of this organisation was changed to MVEE (Military Vehicles and Engineering Establishment).
2. For discussion of the specialised meaning of these terms see Grigg in Chorley and Haggett (1967) *Models in Geography*, pp. 469, 470, 484. Briefly, 'generic' is distinguished

individuals, which had analogues in the same climatic zone recognisable on aerial photographs, and which could also form the basis of a new practical worldwide regionalisation.

MEXE asked the Combined Pools of the Army Emergency Reserve (AER) to test the idea in representative areas from each of the main climatic zones: arid, savanna, humid temperate, and tropical rain-forest. Short studies in the Near and Middle East, East Africa, UK, and Malaya (MEXE, 1965) gave encouraging results and led to a formal project, under Beckett and Webster, to test the idea critically in detail in the UK.

A study was made of a 5000 km^2 area in the English Midlands, roughly centred on Oxford. The first difficulty was that the more exact the definition and the more homogeneous the properties demanded of the proposed land units the harder it became to recognise them on aerial photographs. It was therefore necessary to evolve a compromise whereby approximately equal weight was given to the two considerations, and only to identify units that were normally and easily recognisable.

It was found possible to divide the area into three natural physiographic regions (River valley on clay, chalk, and scarplands) relevant at the same range of mapping scales (1:250 000–1:1M) as had been used in CSIRO studies (see chapter 9). Following earlier work, these were called *recurrent landscape patterns* or RLPs, though since this time this term has been generally replaced by the Australian term *land system*. These were divided into seventeen *facets* of which a map was produced at a scale of 1:63 360 (Beckett and Webster, 1965a).

It was demonstrated that both types of unit recurred and could be recognised and delimited on aerial photographs, with the aid of geological maps and a relatively small amount of field work, and that each was adequately homogeneous and different from others to be used for practical predictions within the study area (Beckett and Webster, 1965b, c, d). The authors have more recently published summaries of this work (Beckett and Webster, 1969; Webster and Beckett 1970).

On the strength of these results, work was initiated at Cambridge with the aid of developing and testing a comprehensive terrain classification over a whole climatic region, the hot arid zone.

An exhaustive study of the literature and a representative sample of air photo cover, followed by a field visit to USA produced about 700 descriptions of locally occurring varieties, called *local forms*, of facets which were then grouped into geomorphological types and a unifying

from 'unique' and implies that the units are recurrent, 'uniform' is distinguished from 'nodal' and implies that they are not 'nucleated' as for instance are urban hinterlands. 'Operationally defined' implies that although the units have a 'natural' basis, they are not entities but are arbitrarily chosen to meet practical needs. Recognising that they are 'individuals' allows them to be treated by normal methods of scientific classification despite the fact that each possesses a unique property in this location.

concept called a *facet abstract* defined for each. As these were aimed to suit a mapping scale of 1 : 100 000, some 'facets' showed so much internal variation in soil and mesorelief that they could not be considered as facets in the sense used hitherto. Accordingly, the term *clump* was introduced for these, bearing an analogous relationship to the facet as does the *complex* to the *series* in soil mapping. Altogether fifty-six facet and forty-two clump abstracts were defined.

About 400 *local forms* of larger units suitable to a mapping scale of about 1 : 1M were then grouped into thirty-nine *land system abstracts* by the same process.

Having demonstrated that a classification could be comprehensive and yet not contain an unmanageable number of facets, the next stage was to test their uniformity. The relevant units were mapped for areas in Libya, Trucial Oman, and the islands of Bahrein, Socotra, and Abdul Kuri. Expeditions were mounted to make field measurements, and take samples. The data were analysed statistically to test facet homogeneity. It was concluded that they were sufficiently homogeneous and mutually distinct to allow valid predictions within a local form of a single land system, but that beyond its limits predictions could only be made if there was close geographical proximity or marked physiographic similarity between the known and the unknown areas.

This finding is important and is supported by the available evidence from other climatic zones. It shows that there is a *prima facie* case for assuming that the world is divided into a finite number of land systems, each with constituent facets which are sufficiently homogeneous to be treated as uniform for moderately extensive land uses. In the military context this means that armies can benefit from a wholesale subdivision of the landscape into land systems and facets. The former will be more useful units at the broader scale of strategic planning at theatre and army corps level, but the latter are preferable over areas of a few miles or a few tens of miles – the scale of operations of the division or battalion. When larger scales than this are required the value of the whole land system facet concept diminishes because of the difficulty of recognising accurately the finest subdivision of the landscape or of communicating this information, once recognised, to the multitudes of individual NCOs and soldiers, each of whom would be confronted with a complex and ever changing series of decisions involving at most a few metres of terrain.

Since the conclusion of this research, development of a terrain prediction capability in the UK has passed into civilian hands to which its military applications are peripheral. A practical system of military terrain intelligence could, however, be derived at relatively short notice if required.

The Indian Army have adopted a similar system (Beckett, 1967) and have classified areas in the Punjab plains, round Sangor (MP), and in Ganhati – Assam, and carried out some studies on the uniformity of terrain units.

A landscape approach has also been employed by the US Air Force Cambridge Research Labaratories (hereinafter referred to as AFCRL) in a series of detailed studies of tropical soils.

The characteristics and airphoto interpretation of tropical soils were reviewed by Ta Liang (1964), including consideration of their origin, physical and chemical characteristics, and the engineering problems likely to be experienced on each of the major soil groups. A key was given by which these groups could be identified on aerial photographs.

Playas were studied by Motts (1970) among others. They occur widely in all arid areas, though the term is derived from south-western USA. They have been defined as level or nearly level areas occupying the lowest part of completely closed basins that are covered with water at irregular intervals forming temporary lakes. They are generally composed of clay or silt containing large amounts of soluble salts. The surface is usually devoid of vegetation and may be hard or soft, smooth or rough. The width and flatness of playa surfaces has important military implications. Apart from containing deposits of exploitable minerals, they can be used seasonally for the landing of aircraft without pretreatment, and they smooth themselves after rain. They provide possible recovery sites for spacecraft or locations for large arrays of antennae. They also provide analogy with expected conditions on the moon and Mars, and for this reason have been used in designing space landing equipment.

As much data as possible has been summarised about a large number of American, Australian, Iranian, and North African playas, especially their setting, surface characteristics, hydrology, physical properties, minerology, tectonic framework, basin geophysics, and characteristics recognizable on aerial photographs. They have been classified in terms of the chemistry and hardness of their surfaces, and special attention paid to the surface stability and giant desiccation polygons which could affect their usability for aircraft. Finally, the potential of satellite photographs was shown in revealing detailed seasonal changes on the surface of certain American examples.

Studies were made to see how far data on playas could be quantitatively assessed from aerial photographs, using a microdensitometer technique. It was found that distinctions between terrain types could be rapidly scanned but that it was not possible to quantify their internal character or to separate the image effects due to vegetation, soil, and cultural factors.

The Cornell Aeronautical Laboratory have carried out a programme of military research which evaluated broad areas of the earth's surface in climatic and physiographic terms in relation to general transport systems requirements. They devised a quantitative system for determining from maps the degree to which topography limits lines of sight range surface targets from a source (Deitchman, undated).

Parametric systems of military terrain evaluation

The parametric approach has been most fully developed by the US and Canadian armies, mainly by the Quartermaster Research and Engineering Centre at Natick, Massachusetts (QREC), the US Army Engineer Waterways Experiment Station at Vicksburg, Mississippi (USAEWES), and the Canadian Defence Research Board. These three organisations differ somewhat in objective, the first being concerned mainly with aspects of the environment from which an army requires protection, and the other two with means of exploiting the environment to military advantage, specifically cross-country mobility.

QREC have produced terrain bibliographies of the Russian Arctic and western Africa, environmental handbooks on different parts of the world, and have also evolved methods for analysing lines of sight and surface relief. Early studies in 1954–56 established three basic points: that average slope varied directly with average relief, that the latter could be more accurately predicted when natural vegetation and geology were known, and that relief over a small area (9 km²) could be predicted from data gathered from maps at too small scales (1:¼M and 1:1M) for such areas to be shown in detail. In order to derive a predictive method for topographic analysis, 204 separate maps of part of USA at a scale of 1:62 500 were chosen at random and ten sample areas selected in each by the expedient of drawing concentric circles round the centre of the maps to enclose respectively $\frac{5}{16}$, $\frac{5}{8}$, 1·25, 2·5, 5, 10, 20, 40, 80, and 160 square miles. For each circled area the following measurements were made: (1) highest elevation, (2) lowest elevation, (3) relief (obtained by subtracting (2) from (1)), (4) number of closed contours ('hilltops'), (5) number of crossings of twenty contours in directions due N–S and E–W from the centre point ('contour count'), and (6) number of valleys and divides on the same traverse as used in (5) ('slope direction changes'). This process yielded six items for ten sample units on each of 204 sheets or 12 240 separate items of data.

When it was seen that the variation was orderly, this number was reduced by ignoring every other sample area, i.e. those enclosing $\frac{5}{16}$, 1·25, 5, 20, and 80 square miles.

A ranking method of analysis was used. Correlations were found in all sized areas between relief, contour counts, and slope direction changes. The mean relief of a $\frac{5}{16}$ square mile area was 240 ft and for a 160 square mile area was 1420 ft. A successive doubling of areas from the smallest to the largest was accompanied by mean values for relief which fell into a regular progression. This permitted determination of approximate mean relief for areas in the USA of any size from $\frac{5}{16}$ to 160 square miles. Certain values were of interest. The average elevation of the USA is about 2300 ft. On a random traverse, a 20 ft contour was crossed every 290 ft along the ground, and three ridges averaging 150 ft above valleys were encountered every mile. Three hilltops sufficiently

81

extensive to be represented by an enclosed 50 ft contour occurred on average in every square mile. These values were means, and the medians were appreciably lower, indicating that the distribution of data was strongly skew (Wood and Snell, 1957, 1959).

Six somewhat different quantitative indices were used in a further study of that part of Europe lying between 48° and 52°N and 7° and 16°E (Wood and Snell, 1960). These were *grain, relief, average elevation-relief ratio, average slope, and slope direction changes*. They were measured on 1 : 100 000 sheets of the US Army Map Service Central Europe Series, having a contour interval of 25 metres, and gave the following results.

Grain was the spacing of major ridges and valleys. It was assessed by selecting a random point on a map, drawing a series of concentric circles having diameter increments of one mile, and determining the maximum difference within each circle. When these values for relief were plotted against increasing sample area size, it was found that a 'knick point' occurred where relief ceased to increase appreciably. The sample area size equivalent to this knick point represented the grain of the area. *Relief* was the difference between the highest and lowest elevations in the unit area equivalent to the grain size.

Average elevation was derived from the mean of nine randomly chosen points within the unit area. *Elevation-relief ratio*, the relative proportion of upland and lowland, was derived by subtracting the lowest elevation from the average elevation within the area and dividing the remainder by the relief. The resulting value therefore always fell between 0 and 1 and was expressed as a decimal. *Average slope* was determined by counting the number of contours crossed by straight lines in directions NW–SE, N–S, NE–SW, and E–W across a circular unit area equivalent to the grain size, and computing the slope tangent from the equation

$$S \tan = \frac{I \times M}{3361}$$

where : S tan is the slope tangent
$\quad\quad\quad$ I is the contour interval in feet
$\quad\quad\quad$ M is the number of contours crossed per mile of random traverse.

Slope direction changes denoted the dissection of an area and were another expression of topographic texture. They were assessed by counting the number of changes from rise to fall and vice versa along the same random traverses used for counting contours.

When the results of this study were analysed, it was found that groupings of the numerical data from different indices gave twenty-five distinct regions which conformed well to the landform regions identified on previous physiographic maps, such as those of Lobeck (1923) and van Valkenburg and Huntington (1935) and had the double advantage of being based on an essentially simple analysis and resulting in groups which were quantitatively defined.

The 'Vicksburg' approach

The US Army Engineer Waterways Experiment Station at Vicksburg, Mississippi, is responsible for evaluating the effects of terrain on military activities, and carried out a programme of research, inspired by Dr Paul Siple in 1953 and called MEGA (Military Evaluation of Geographic Areas). This began with a selection of the key terrain factors on which study should be concentrated. These had to be based on the following overriding needs: to limit the total number of factors to manageable proportions, and to favour those basic elements in the terrain which were easily visualised, simple to map, significant to military uses, suitable in developing analogues between different areas, and which together gave a complete picture of the terrain.

Numerical subdivisions of the parameters so derived had to be suitable as mapping units. Careful consideration had therefore to be given to the availability of data, especially in the less well mapped parts of the world, to their military significance and their amenability to quantitative measurement. Natural breaks in the landscape were used wherever possible. It was found, for instance, that alluvial fans almost always sloped less than 6° and the windward slopes of barkhans between 5° and 14°. This permitted 6° and 14° to be used as critical values.

These principles led to the selection of certain terrain factors in an initial study of deserts, which fall into groups, called 'factor families', whose constituents and internal classes are considered in detail in a handbook (USAEWES, 1959).

The first of these factor families were called *aggregate and general factors*, consisting of *physiography, hypsometry,* and *landform and surface conditions*. Physiography was a useful generalisation of surface geometry factors. The terms 'plateau', 'plain', and 'hill' were included for the usefulness of the generalisation rather than the value of the measurements. Hypsometry showed altitude classes. As vehicles show efficiency breaks at approximately 5000 ft and 9000 ft, these two limiting contours were included in the classification. Landform and surface conditions were shown by physiographic sketch-maps by Raisz and others following his style (Raisz, 1938, 1946), and by outline maps of the main geomorphic surface types: depositional, erosional, tectonic, volcanic, intrusive, etc.

Each of the surface geometry or form factors was assessed quantitatively. *Characteristic slope* was the narrow range of slopes which were commonest in the region under consideration as revealed by the 10 ft contour interval. The class divisions used aimed at being natural and accorded with research which had shown that the tangents of observed · slope angles tended to cluster mainly round values of 0·05, 0·1, 0·2, 0·4, 0·6, and 0·8. *Characteristic relief* was the maximum difference in elevation per unit area. This had to be assessed differently depending on whether the characteristic slope of an area was greater or less than 6° and whether the drainage lines were well or poorly developed. *Occur-*

rence of slopes greater than 50 per cent was assessed by counting the frequency that such slopes occurred on sample transect lines. *Characteristic plan-profile* really defined the areal relations of the three preceding factors. It attempted to express the 'peakedness', areal occupance, degree of elongation, and the orientation of topographic highs in quantitative terms to form a legend. The classes derived are as shown on Fig. 10.2. They total twenty-five as each of the four kinds of plan arrangement

Plan arrangements:

1. Non-linear and random.

2. Linear and random.

3. Non-linear and parallel

4. Linear and parallel.

Profile arrangements:

Flat topped (summit area slopes $<6°$, side slopes $>14°$)

1. Highs occupy >60 per cent of area.

2. Highs occupy 40–60 per cent of area.

3. Highs occupy <40 per cent of area.

Crested or peaked (limiting slopes $>6°$)

4. Highs occupy >60 per cent of area.

5. Highs occupy 40–60 per cent of area.

6. Highs occupy <40 per cent of area.

7. No pronounced highs or lows.

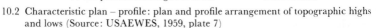

10.2 Characteristic plan – profile: plan and profile arrangement of topographic highs and lows (Source: USAEWES, 1959, plate 7)

can be combined with each of the six kinds of profile. The only other class is that without pronounced highs or lows which for this reason has no particular plan arrangement. All the different factors in the surface geometry family can be synthesised into a map of generalised landscape although care must be taken to avoid a plethora of mapping units if too many different combinations of individual factors occur.

The *ground and vegetation* factor family consisted of five properties. *Soil type* generally related to soil texture and proportion of surface occupied by bare rock. *Soil consistency* depended on the degree of layering, cohesiveness, and crustness in the soil. *Surface rock* supplemented soil type by showing the lithology of the rock which was exposed at the surface, and *vegetation* used a quantified physiognomic classification based on Dansereau (see chapter 7). Although the importance of microrelief, arbitrarily defined as that with a vertical amplitude of less than 10 ft, was recognised, it was not included in the scheme because of the necessity of imposing a lower limit on the scale of generalisation and the fact that there is a great lack of information on the surface of the earth at this degree of detail.

All factors excluding microgeometry were investigated and maps were produced, in the first instance, of world deserts at a scale of approximately 1:5M. These maps aimed at being data sources which could be used to answer specific questions. For instance, trafficability questions would demand overlaying the maps of soil strength, characteristic slope, slope occurrence, and possibly others, to obtain composite parametric units. A search for sources of building materials would only require the use of the surface rock and soil type maps, and so on.

The Yuma Test Station, Arizona, was mapped with the same factors at a scale of 1:400 000. The purpose of this was to determine the degree of anology between Yuma and other world deserts, to provide the basis for comparing world deserts with each other and with Yuma, and to help evaluate the relations between terrain and such quantified military activities as the trafficability of given vehicles or the grading of naturally occurring materials required for military construction. To achieve the first purposes 'analog maps' were produced which assessed the world deserts in terms of their degree of similarity to Yuma.

More recently, other areas have been mapped using the same parameters (e.g. Fort Leonard Wood, Missouri), and analyses made of the relations between vehicle mobility and soil strength, surface geometry, and tree stem spacings in test areas (USAEWES, 1963). The air photo recognisability of the quantified classes was explored, but it was concluded that only the physiographic landscape types were clearly recognisable.

Finally, an attempt was made to analyse terrain by electromagnetic means (USAEWES, 1965–67). This explored the possibility of recognising certain terrain features by controlled experiments in four different spectral regions, using sensors held at fixed heights about the ground.

Investigation of the 0·76–5 micron spectral band using electromagnetic sensors showed that the type and moisture content of the soil could be determined under controlled laboratory conditions. Radar was able, within strict limits, to determine the moisture content and depths of soil layers. The 0–2·82 MEV gamma spectral region had some capability in determining soil moisture, and an attempt was made to determine if the soil type, moisture, and dry density could be determined using infrared radiation below a wave length of 7·7 microns. Although a relation between emissivity and soil surface moisture appeared to exist, it was so influenced by surface temperature and incident radiation that

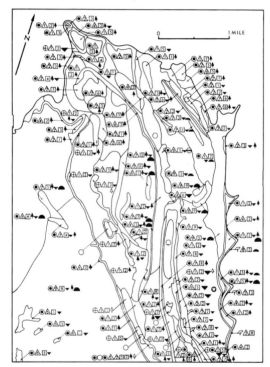

Composite terrain map of part of Camp Petawawa, Ontario

LEGEND FOR TERRAIN MAP

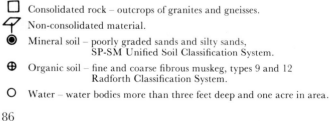

Surface Composition

☐ Consolidated rock – outcrops of granites and gneisses.

Non-consolidated material.

◉ Mineral soil – poorly graded sands and silty sands,
   SP-SM Unified Soil Classification System.

⊕ Organic soil – fine and coarse fibrous muskeg, types 9 and 12
   Radforth Classification System.

○ Water – water bodies more than three feet deep and one acre in area.

Surface Morphology: Macromorphology

Slope steepness:

⚠1	0 – 6°	0– 10%
⚠2	6–14°	10– 25%
⚠3	14 –26½°	25– 50%
⚠4	26½–45°	50–100%
⚠5	Above 45°	Above 100%
⚠6	Classes I and II	
⚠7	Classes II and III	
⚠8	Classes III and IV	
⚠9	Classes I, II, and III	

Slope form:

1	Convex, smooth
2	Planar, smooth
3	Concave, smooth
4	Convex, rough
5	Planar, rough
6	Concave, rough
7	Classes 1 and 3
8	Classes, 1, 2, and 3
9	Classes 4, 5, and 6

Surface Morphology: Micromorphology

● Positive features of mineral soil in a random, linear pattern. Slopes 6   7 , lengths 400–1800 ft, width–length ratio 1:4–1:20, amplitude 10–30 ft, spacing 10 per mile, non-symmetric sigmoid in section. Aeolian – fixed sand dunes.

● Positive features of mineral soil in a random, linear pattern. Slopes 6   7 , lengths 50–400 ft, width–length ratio 1:2–1:8, amplitude less than 10 ft, spacing 18 per mile, irregular sigmoid in section. Aeolian – sand sheets and ripples.

⊖ * Negative features in mineral soil in a clustered, non-linear, overlapping pattern. Slopes 7   9 , lengths 10–200 ft, width–length ratio 1:1–1:2, amplitude 10–50 ft, spacing 15 per mile, irregular cardioid in section. Glaciofluvial – kettle holes.

● * Positive features of consolidated rock in a random, non-linear pattern. Slopes 5   5 , lengths 20–100 ft, width–length ratio 1:1–1:2, amplitude 10–30 ft, spacing caculation not possible, irregular rectilinear in section. Glacial – rock outcrops and erratics.

* These symbols are not included on map.

Surface Cover: Vegetation Structure

	Height:	Stem type:	Form:
↑	More than 25 ft	Woody	Trees
⇡	5–25 ft	Woody	Young or dwarfed trees
▼	2–5 ft	Woody	Tall shrubs or dwarfed trees
▼	Less than 2 ft	Woody and non-woody	Low shrubs, grasses, sedges, and mosses

Surface Cover: Vegetation Spacing (mean nearest neighbour distance[8]

⑨	0–10 ft	④	60–90 ft
⑧	10–15 ft	③	90–140 ft
⑦	15–25 ft	②	140–220 ft
⑥	25–40 ft	①	Greater than 220 ft
⑤	40–60 ft		

10.3 Example of the system of terrain mapping employed by the Canadian army. (Source: J. T. Parry, J. A. Heginbottom and W. R. Cowan, 1968, pp. 163, 165)

it did not appear a generally usable method in predicting soil trafficability values.

More recently, the Vicksburg work has moved far in the direction of automation (Grabau, as quoted by Grant, 1968). Research is no longer devoted to terrain evaluation *per se* but rather to the development of mathematical models of all conceivable military engineering activities involving terrain as a component. To this end engineer laboratories have begun to abstract from the literature any engineering formulations that might be adapted to military purposes, such as standard design for highways and airports, structural relations in bridges, etc. The resulting body of mathematical knowledge will be used to identify the parameters that the military will require in the future, and initiate the development of the instrumentation by which they can be acquired. There is already enough need for this ground information to justify automation wherever possible. For example, a laser theodolite can automatically measure horizontal and vertical angles and put the information directly on to magnetic tape which a computer can be programmed to print out as a topographic profile (*ibid*).

Canadian army system

The system employed by the Canadian army is mainly concerned with vehicle mobility. Parry, Heginbottom, and Cowan (1968) describe how Camp Petawawa, Ontario, was used as a sample area and surveyed in detail to determine the potential of large scale photographs (1 : 5000) in assessing the environmental factors affecting the cross-country mobility of military vehicles. The parameters considered were: surface composition, macromorphology, micromorphology, and surface cover. Map overlays were prepared of each, and a composite map produced, part of which is shown on Fig. 10.3. Surface composition was based on the rigidity, elasticity, and viscosity of the surface materials. Macromorphology consisted of a subdivision of slope into steepness classes, distinguished according to whether they were convex, planar, or concave. Micromorphology was found to be the most difficult factor to assess on aerial photographs because it consisted of a detailed evaluation of the composition, organisation, dimension, form, and genesis of small surface features. Surface cover consisted in a quantification of man-made and vegetation features. Tests in the area indicated that vehicle performance was similar for areas with similar arrays of terrain characteristics, and on this basis successful predictions of speed were made for cross-country test runs traversing a variety of terrain units.

On the whole, the military evaluation of terrain is still in its infancy. The most likely future trend is towards an increasingly sophisticated measurement of landscape parameters combined with more detailed examination of their effect on military activities, aiming always at a greater capability in reacting to, managing, and modifying the environment.

11
Terrain evaluation systems for engineering

The difficulties encountered in providing terrain data for engineers derive from the wide scope of engineering activity, the variety of its points of contact with terrain, the differences in the degree of detail and duration of individual projects, and the intangibility of many of the relationships between engineering works and the terrain they use.

There are basically two types of environmental data needed by engineers: the tangible and the intangible. Within the first type is information concerning the *form* of terrain, such as the dimensions of natural features, and the locations, amounts, and characteristics of specific materials. Within the second is information about geomorphological, pedological, and hydrological *processes*, and engineers' experiences of dissimilar, but related areas, especially when these include impressions of success and failure. The basic problem is to devise some means of saving both tangible and intangible information in such a way that it can be readily stored and communicated.

On the whole, most progress has been made in those branches of engineering which depend on terrain factors most directly and extensively, and in those in which terrain factors govern the highest proportion of total cost. These are mainly the military, hydrological, and highway aspects. The first two are considered in chapters 10 and 12 respectively; the third is considered here.

Although there are differences in practice, highway engineering is taken to include not only roads, but also railways and airfields. The highway engineer has always faced a number of constraints in selecting a route. First, he must minimise the distances between the fixed end points of the road. Secondly, he must minimise gradients: these cause construction and maintenance costs to multiply in direct relation to their steepness because they necessitate greater lengths, larger amounts of cut and fill, additional construction works such as spirals and bridges, and protection against landslips and other forms of erosion. Thirdly, he must avoid watercourses as far as possible. Langbein and Hoyt (1959) estimated that no less than one-quarter of the capital expenditure on roads in USA was for the construction of bridges, culverts, and other drainage works. Fourthly, the engineer must choose a route which will not only balance cut and fill but will traverse areas where construction materials of suitable nature and particle size composition are available. Of recent years a fifth factor has entered into consideration: the soil. If only the other constraints are observed it is still possible to incur serious trouble by traversing areas of unstable soils, especially if they are composed of silt, clay or organic materials.

The catalogue of terrain factors important to highway engineering is long. At the planning level, it begins with the geological structure and the general climatic environment, notably the rainfall isohyets. At the project level, it includes the location, areal extent, and thickness of rocks, soils, and gravel deposits; the volume, distribution, and quality of water supplies, the depth to the water table both in the wet and dry seasons, and the general drainage situation. It also includes information about surface materials, such as their rippability, cut-slope stability, swelling and shrinkage characteristics, compactibility, and firmness in all moisture conditions (expressed as California Bearing Ratio or CBR). At a finer scale, it involves knowledge of the texture and Atterberg limits of soils. Finally, special site information is often required for such things as the construction of bridges and culverts.

The problem is essentially to use information to harmonise objectives and overcome constraints in the cheapest possible way, an operation which can be most efficiently done with the aid of computers.

The most important terrain evaluation systems used for engineering have all been developed by road research organisations and are basically of the landscape type. Three are notable: those of the UK Road Research Laboratory (RRL), the South African Department of Scientific and Industrial Research, National Institute of Road Research (NIRR), and the Australian CSIRO Division of Soil Mechanics. Their views have converged, and since the Oxford Symposium (Brink *et al.*, 1965), there has been a certain amount of unofficial standardisation among them and with MEXE and the CSIRO Division of Land Research and Regional Survey (LRRS). The work of these organisations has been summarised by Beckett and Webster (1969).

Road Research Laboratory, UK, (RRL)

The Tropical Section of RRL has studied the occurrence of road building materials in relation to geology, climate, and topography since about 1960. The work started with regional surveys but progressed to more detailed examination of selected areas using aerial photographs to locate road lines, bridge sites, and suitable road-making materials such as laterite gravels. The first detailed terrain evaluation study was made in Nigeria, and subsequent research has been done in Malaya and elsewhere. Since the Oxford Conference (Brink *et al.*, 1965), however, RRL have been using the generally accepted Australian-British-South African land system approach, modified only to suit their particular data storage requirements. A recent example of the method is the diagram of the Doto land system, Nigeria, shown in Fig. 11.1.

National Institute of Road Research, South Africa (NIRR)

In order to design highways which achieved optimal route alignment consistent with accessibility to construction materials, NIRR developed a system of 'Geotechnical mapping' based on aerial photographic interpretation followed by field sampling. The earliest examples were for the Mariental-Asab route and the Etosha Pan area in South West Africa (Kantey and Templer, 1959; Mountain, 1964). The maps showed lithological classes but did not incorporate surface geometry into the definition of units or provide for extrapolation outside the areas studied.

The Symposia at La Hermosa and Magaliesburg, Transvaal (South African Institute of Civil Engineers, 1964, 1965) marked the acceptance in South Africa of the idea of a unified terrain intelligence system based on the land system approach, which, following the Oxford Conference, came into line with practise in UK and Australia.

Since this time, NIRR have aimed to cover southern Africa with higher land units, starting with maps of considerable areas in Transvaal and South West Africa. An index map has been produced, showing the location of specific road building projects and indicating the sources of data about them. Detailed land system mapping is being undertaken in the Johannesburg-Pretoria district and the whole Karoo formation area. An example of a land system from this area is shown on Fig. 11.2.

The distinct emphasis of the South African work is on the short term engineering project level approach. This is because, for any given area, they envisage the production not only of a land system map but also of a soil engineering map which translates the facets into road engineering terms and enables the user to tap the data storage system which is now in operation at NIRR in Pretoria (Brink *et al.*, 1968).

Division of Soil Mechanics, CSIRO, Australia (SM)

SM is a separate section of CSIRO from LRRS and is concerned with the engineering rather than the agricultural aspects of soils. Their mutual contacts in the terrain evaluation field date from 1963 when LRRS requested SM to undertake an engineering assessment of the Tipperary land system in the Katherine-Darwin region, Northern Territory, which the former had previously surveyed. The results of this assessment are contained in a report which evaluates each unit in terms of its engineering characteristics (Grant and Aitchison, 1965). This report included a quantitative tabulation of vegetation height and density dimensions and gradients of slopes, channels, and drainage net, characteristics of the rock outcrops and such considerations as visibility restriction by vegetation and sources and properties of engineering materials.

The inadequacy of conventional land system surveys for engineering purposes soon compelled SM to devise a modified system for their own needs. This had more narrowly defined land systems and facets, which were proved by engineering experiment and were defined entirely on recognition characteristics before specific engineering data were attached to them.

The terrain evaluation process was seen as consisting of three phases: the establishment phase, the quantification phase, and the interpretation and application phase. The first two of these are outlined below, but the third is more appropriately considered with data management in general (chapter 19). The establishment phase was that of terrain classification and mapping in accordance with the rigorous principles applicable to engineering needs. It also involved the establishment of full communication between classifier and user so that the latter could, without training in geomorphology or soils, not only use the system at any level of generalisation but also build his requirements into the original classification.

The proposed engineering terrain classification scheme was based entirely on land properties normally available to engineers and avoided abstractions at all levels. There were four levels of generalisation: *province, terrain pattern, terrain unit,* and *terrain component* in descending order of size. Only the last three were considered as vehicles for engineering data. The method of relating such data to these was called the PUCE (Pattern Unit Component Evaluation) programme of terrain evaluation for engineering, and is described by Aitchison and Grant (1967, 1968a, 1968b).

DOTO LAND SYSTEM

Climate—Semi-arid equatorial tropical (Sudan savanna zone): 3 humid months (July–September), 750–850 mm rainfall. Average daily maximum temperature above 33·5°C.

Rock—Medium- to coarse-grained sandstones of the Paleocene Kerri-Kerri Formation with subordinate conglomerates, clays, and siltstones; flat lying strata with an overall gentle dip to the north. Overlain by a thick, extensive lateritic ironstone which dips beneath Quaternary sediments of the Chad Formation.

Landscape—Extensive dissection has given rise to an intricate system of deep, steep-sided, flat-bottomed valleys, with tabular interfluves and residual mesas, frequently capped with lateritic ironstone. The form of the drainage pattern shows that the former stream courses were part of a system which flowed north-east to Lake Chad but has subsequently been captured by a tributary of the R. Gongola (R. Gaji) with a reversal of drainage. The accumulation of sediments on valley floors demonstrates that active down-cutting has now ceased. Five facets are identified in this land system.

Soils—Thin, stony soils on lateritic ironstone and sandstone. Sandy colluvium and alluvium of flat-floored valley bottoms.
Vegetation—Sudan savanna zone; with mixed *Detarium* woodland on stony upper slopes. Savanna 'parkland' on cultivated valley floors.
Altitude—500 m (approx.)
Relief—100 m.
Reference—Klinkenberg, K. *et al. Soil Survey Section Bull.* No. 21, 1963. Regional Research Station. Ministry of Agriculture, Samaru, N. Nigeria.

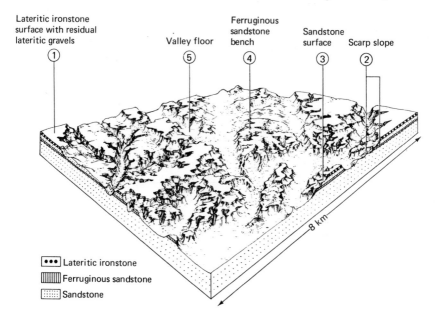

Lateritic ironstone surface with residual lateritic gravels ① | Valley floor ⑤ | Ferruginous sandstone bench ④ | Sandstone surface ③ | Scarp slope ②

8 km

●●● Lateritic ironstone

|||||| Ferruginous sandstone

∷∷∷ Sandstone

Land Facet	Form	Soils, Materials, and Hydrology	Land Cover
1	**LATERITIC IRONSTONE SURFACE** Flat to gently sloping surface at 600 m OD. Occurrences vary in size from 100m² to 2km². Occurs in flat interfluves and isolated mesa surfaces	Lateritic ironstone 8 cm thick occurs as surface capping to sandstone, the upper part of which is often heavily ferruginised. Surface materials comrpise bare ironstone and thin stony soils. Where the laterite has been stripped away, stony soils overlie sandstone	Mixed *Detarium* woodland where sufficient soil occurs
2	**SCARP SLOPE** Uneven steep topography generally becoming steeper on upper part of slope below margin of land facet 1. Concave lower slopes with uneven bouldery microrelief. Steep rocky ribs and pinnacles and incised gullies are common	Lateritic ironstone and sandstone rock, talus and thin, stony soils subject to erosion by hill wash	*Detarium* woodland where sufficient soil occurs
3	**SANDSTONE SURFACE** Flat to moderately sloping ground corresponding to a structural bench formed by dissection of a flat-bedded sandstone	Rock and thin, stony soils	As above
4	**FERRUGINOUS SANDSTONE BENCH** As above, but more persistently developed and determined by the presence of a ferruginous sandstone horizon 0.5m thick, midway between land facets 1 and 5	Rock and thin stony soils. The indurated, ferruginous sandstone protects the underlying softer sandstone and forms a prominent bench, often bare rock, but sometimes overlain by rubbly soils	
5	**VALLEY FLOOR** Flat but steepening slightly at margin with land facet 2. Vary in width from 10m to 900m. Length may be as much as 13 m	Moderately deep, light grey to reddish brown sandy colluvial and alluvial soils. Permeable with no well-defined water courses. Active down-cutting of the gullies has now ceased	Savanna 'parkland' heavily cultivated

11.1 Example of terrain evaluation by the Road Research Laboratory: evaluation of Doto Land System, Nigeria (Source: J. W. Dowling, 1968, pp. 154, 155)

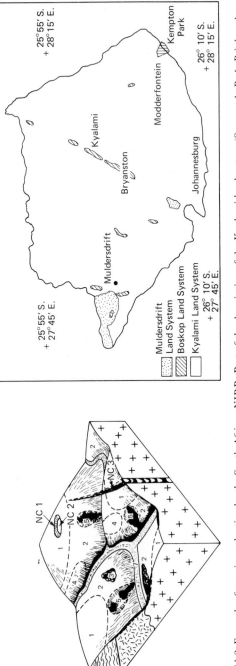

11.2 Example of terrain evaluation by the South African NIRR. Part of the description of the Kyalami land system (Source: A. B. A. Brink and T. C. Partridge, 1967, pp. 10–13.)

TABLE 1 *Kyalami Land System: Table of Land Facets*

Land Facet	Form	Soils, materials and hydrology	Land cover
1	Hill crest. Slope <2° width <2,000 yd	*Variant 1*: (Old erosion surface). Residual sandy clay with collapsing grain structure (<50 ft) on granite. Sometimes overlain by reworked soil (<10 ft) Above ground water influence except at depth. *Variant 2*: Residual expansive clay (<20 ft) on schists and basic metamorphic rocks, with occasional low outcrop. Above ground water influence. *Variant 3*: Weathered granite (<10 ft) sometimes covered by thin vein quartz gravel (<3 ft) and/or reworked soil (<4 ft). Above ground water influence.	Crops and grass
2	Convex side slope (<12° width <1,000 yd).	*Variant 1*: Hillwash of silty sand derived from granite (<3 ft) on granite, schists or basic metamorphic rocks. Occasionally saturated. *Variant 2*: Hillwash of expansive silty clay derived from schists or basic metamorphic rocks (<3 ft), on granite, schists or basic metamorphic rocks. Occasionally saturated.	Crops, grass and low bush.
3	Tor. Slope <90° diameter <100 yd	Fresh granite outcrop on side slope. Above ground water influence.	Trees and bush
4	Whaleback. Slope <10° width <200 yd	Fresh granite outcrop on side slope. Above ground water influence.	
5	Gully. Slope <15° width <300 yd	*Variant 1*: Sandy gully wash derived from granite (<4 ft) on granite. Periodically saturated and sometimes containing ferricrete with perched water table. *Variant 2*: Expansive clayey gully wash derived from schists and basic metamorphic rocks (<10 ft) on granite, schists and basic metamorphic rocks.	Bush, grass and crops.
6	Alluvial terrace. Slope <10° width <100 yd, with steep bank overlooking stream (height <25 ft slope <30°).	Sub-angular gravel and boulders of mixed origin in silty matric (<20 ft) on granite, schists and basic metamorphic rocks. Above ground water influence.	Grass and trees.
7	Alluvial floodplain. Slope <10° width <100 yd with incised stream course flanked by vertical banks of <15 ft	Expansive alluvial clays and sands (<20 ft) on granite schists or basic metamorphic rocks. High water table.	Grass.
Non-cognate facet			
1	Pan side slope (<20° width <100 yd).	Hillwash of silty sand derived from granite (<5 ft) on granite. Occasionally saturated near foot of slope.	Bush and grass.
2	Pan floor. Slope <1° diameter <200 yd.	Poorly drained black expansive clay (<3 ft) on granite. High water table.	Grass.
3	Dyke ridge. Height <20 ft slope <30° width <150 yd.	*Variant 1*: Bouldery outcrop of diabase syenite or felsite. Usually above ground water influence.	Trees and bush.
4	Pediment (below marginal escarpments) slope <9° width <2,000 yd.	*Variant 2*: Residual expansive clay (<10 ft) on diabase, syenite or felsite. Usually above ground water influence. Hillwash of silty sand derived from quartzite (<5 ft) on granite, schists or basic metamorphic rocks. Usually above ground water influence.	Crops and grass.

11.2 (continued)

95

TABLE 2 Kyalami Land System : Elements of the Photographic Image

Facet	Tone	Texture	Shape and internal pattern*		Stereoscopic appearance	Associative characteristics
			Shape and structural pattern	Drainage form*		
1	Light grey.	Moderately fine.	Amorphous, showing occasional angular jointing pattern in igneous rocks, and foliation lines in schists.	Absent except on old erosion surface where dislocated, with occasional pans.	Gently sloping, convex.	Some agriculture.
2	Medium grey with occasional dark patches	Medium, sometimes mottled.	Amorphous around hill crests, no S.P.	Sub-parallel/radial, low density, good integration	Gently sloping, convex.	Patches of low bush and crops.
3	Light grey with heavy shadows.	Medium.	Circular, with angular jointing pattern.	—	Steep, with substantial relief.	Quarries.
4	Almost white.	Moderately fine.	Amorphous showing occasional angular jointing pattern.	—	Gently sloping, convex.	Absence of vegetation.
5	Medium grey, sometimes patchy	Fine, except where rilled at head.	Hemi-lemniscate, no S.P.	Centripetal/sub-parallel at head. Dense but poorly integrated.	Gently sloping, concave.	Tall grass, crops, sand quarries.
6	Light grey.	Medium.	Linear/lenticular, no S.P.	—	Gently sloping, with steepside.	Trees.
7	Medium grey.	Fine, except where gullied.	Linear, sinuous, no S.P.	Sub-parallel when gullied. Moderately dense but poorly integrated.	Gently sloping, except where incised by stream.	Dams, tall grass.
N.C. 1	Medium grey with occasional dark patches	Medium, sometimes mottled.	Ring-shaped, no S.P.	—	Moderate slope, concave.	Patches of low bush, sand quarries.
2	Medium grey.	Very even, fine.	Circular, no S.P.	—	Virtually flat.	Standing water, beach line, tall grass.
3	Medium grey to black.	Mottled, moderately coarse.	Linear, with occasional angular jointing pattern.	—	Broken low ridge.	Trees.
4	Medium grey.	Medium.	Broad belt below marginal escarpments, no S.P.	Sub-parallel, low density, good integration.	Gently sloping, concave.	Crops.

*Includes pattern, density and degree of integration.

11.2 (continued)

The key subdivision was the terrain unit which had to be readily recognisable and was defined as 'an area occupied by a single physiographic feature formed of a characteristic association of earthen materials with a characteristic vegetative cover'. It was mapped at a scale of 1:50 000 and described qualitatively in terms of principal soil, rock, and vegetation characteristics, and quantitatively in terms of lateral dimensions and relief using either aerial photographic interpretation or ground methods.

The smallest unit was the terrain component, which had a constant rate of change of slope, consistent soil at primary profile level, and consistent vegetation associations. It was generally too small to be mapped or interpreted on aerial photographs. It was mapped *in situ* and defined in terms of dimensions of physiography, rock, soil, and vegetation, and of its relative importance within the terrain unit.

The terrain pattern was recognized mainly from its appearance on aerial photographs. It had constant geomorphology and a constant association of terrain units. It was mapped at 1:250 000 and represented on an illustrative block diagram (see Fig. 11.3), and defined qualitatively in terms of its principal soil, rock, and vegetation characteristics and quantitatively in terms of its lateral and vertical dimensions.

The province was also mapped at 1:250 000 but related only to areas of constant geology as revealed on aerial photographs.

The SM scheme, therefore, differs little from others of the same type. Whereas the MEXE–NIRR system includes a geomorphological specialist at all stages in the operation of the classification, the SM scheme, following its initial establishment, depends entirely on the engineer. Moreover, SM differs from LRRS in attempting extrapolations between terrain units in a analogous terrain pattern and differs from MEXE or NIRR in using vegetation characteristics as definitive of classes.

The quantification phase follows the establishment phase, and consists in the attachment of engineering information to the different terrain classes. Because of the range and complexity of such information, it must normally be handled by a comprehensive system of data storage and manipulation.

Quantification techniques have been better developed for physiography than for quantifying engineering information mainly because of the number and complexity of the latter when all forms of an engineering are considered. Most of the detailed work of this type is carried out by military bodies concerned with the cross-country mobility of vehicles, which have been considered in chapter 10.

Province No. 43.001 Rolling Downs Group

TERRAIN PATTERN No. 04

LITHOLOGY — Shale, claystone, siltstone, sandstone; often overlain by tertiary or quaternary sandstone, conglomerate, gravel, clay, silty clay, often gypsiferous

OCCURRENCE — Scattered occurrences mostly bordering drainage systems adjacent to Flinders and Willouran Ranges

TOPOGRAPHY — Moderately undulating dissected terrain

INCLUSIONS — Terrain patterns 01/3, 11, province 43.001, terrain pattern 01, province 50.007

NOTE - Parts of this terrain pattern may be covered by a veneer of aeolian sand

CHARACTERISTIC CROSS-SECTION SHOWING TYPICAL LOCATION OF TERRAIN UNITS

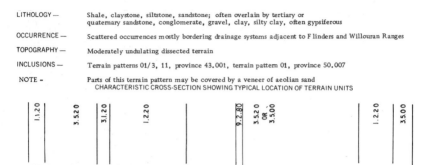

Vert. scale = twice horiz. scale

0   0.25   0.50

Miles

TERRAIN UNITS

Number	Terrain Pattern Area (%)	Occurrence	Description of Dominant		
			Topography	Soil	Surface Cover
1.1.20	5	Alternate to terrain unit 1.2.20	Flat surface	Stratified yellow brown medium to heavy-textured sandy clay with lenses of gravelly clay (CL-SC-GC), over variable gypsiferous sands and gravels (SC-GC) or occasionally over grey or purple decomposed and/or wholly or partly silicified rock of various ages	Sparse to mediumly dense rounded silcrete and quartz
1.1.80	5	Adjacent to drainage below terrain unit 3.5.20	Flat surface (floodplain)	Stratified medium-textured sandy or gravelly clay (CL-SC-GC) over red brown heavy-textured clay (CH) with lenses of sand and gravel	Mostly nil; occasional areas of sparse rounded silcrete
1.2.80	30	Continuous; extensive; included all other terrain units	Gently undulating surface	Stratified red brown medium to heavy-textured clay (CL-CH) over sandy or gravelly clay (SC-GC) over highly gypsiferous clay commonly with kopi lenses (ML), over grey, or purple decomposed and/or wholly or partly silicified rock of various ages	Sparse to mediumly dense rounded silcrete and * quartz
3.1.20	<1	Discontinuous; mostly replacing terrain unit 3.5.00	Smooth steep slope	Grey gypsiferous silt over heavy-textured clay (decomposed rock) (ML/CH); some areas may have a capping of wholly or partly silicified purple sandstone and conglomerate	Mostly crystalline gypsum and kopi rubble, areas of rubble derived from rock outcrop

TYPICAL DRAINAGE NET OF TERRAIN PATTERN

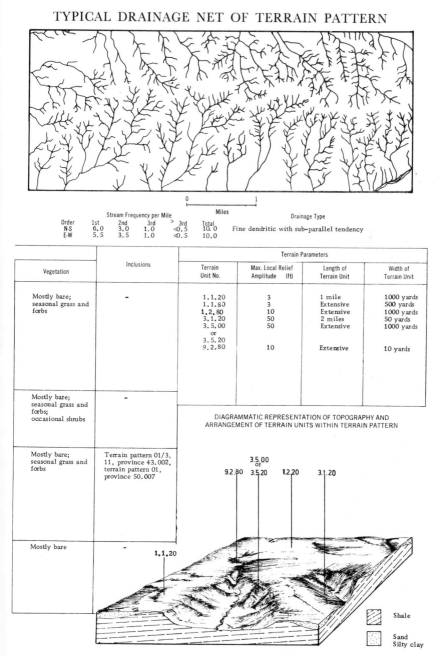

11.3 Example of terrain evaluation by the Australian CSIRO Division of Soil
Mechanics: Terrain pattern No. 04 (Source: K. Grant, 1970, pp. 52–3)

12
Landscape analysis in meteorology and climatology

Terrain classification has seldom been applied directly to problems of meteorology or climatology. This is partly because the relationships are complex and difficult to define, and partly because they have hitherto appeared less economically important than have the connections between terrain and, say, engineering or agriculture.

Nevertheless, land values and uses even on a local scale are fundamentally affected by factors that are basically climatic. Agriculture is dependent on soil and terrain conditions which are in turn governed by seasonal patterns of heat, light, and rainfall regime. Residential and recreational land values are higher on sheltered south-facing sites in the north temperate zone, but on higher and more exposed sites in the tropics.

Topography and local climate

The main effects of macrorelief on local climate are well known: the increasing solar radiation, rain, winds, and cold with altitude, the interference with large air masses to cause high rainfall on windward slopes and dry down-valley winds of foehn type on leeward slopes, the differentiation between sites on the basis of aspect and exposure, the generally small annual and daily temperature variations, the importance of size, direction, and configuration of valleys and basins in determining local air movements, and the delayed temperature maxima and minima especially when the melting of snow and ice is involved. Relatively short distances see sharp changes in climate.

Given quiet general weather conditions, e.g. under anticyclones, topographic factors have their greatest effect on microclimate in the daytime because of the difference of insolation received by different sites. The contrasts are most marked in valleys and depressions, which have quite different diurnal thermal regimes from their surrounding slopes. Fig. 12.1, representing a normal valley, illustrates this. In the daytime, temperature conditions are fairly uniform and ground values depend mainly on aspect. During the course of the day, there is a tendency for the air in the bottom of depressions to become warmer and for air currents to move up the side slopes. At night this condition is reversed and cool air, being heavier, tends to move downslope into the depressions, and form cold air or frost pockets in which condensation often occurs, causing mist or fog.

There are several reasons for this inversion and the development of low temperatures in dish-shaped hollows. There is a downward drainage of cooler, heavier air, and a reduction of turbulent flow in the bottoms. There is a shorter time between sunrise and sunset during which insolation can take place, and a relatively longer time during which outward reradiation can take place. The speed of formation of such cold air pockets is inversely related to their size.

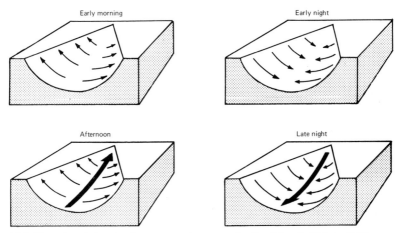

12.1 Diurnal variations in air movement in a normal valley (after R. Geiger, 1965)

When the hollow is an elongated valley rather than a closed depression, the situation is somewhat less simple. In the daytime, after the warmed air has begun to rise up the valley slopes, a second movement of air takes place, up the valley itself. Similarly, at night a down-valley airflow occurs after the general downslope movements have begun. Temperatures in the valley do not fall so low when the depression or valley is steep and narrow than when it is gentle and wide because there is a greater volume of downflowing air relative to the size of the basin and this air has been more exposed to warming on the slopes during the daytime.

The downward movement of air at night is not a phenomenon confined to hollows, but also occurs on the slopes of isolated hills. In the Sudan, for instance, cool night breezes formed in this way are welcome relief after the heat of the day, and both villages and government rest houses are often sited at the foot of inselbergs to take advantage of it, e.g. Jebel Bozi.

Following these considerations, it is in general possible to discern a threefold subdivision of topographic situations in accordance with diurnal temperature variations, illustrated in Fig. 12.2.

So marked is the contrast between 'warm convex' and 'cold concave' land surfaces that the hollows may at night have temperatures many degrees colder than stations thousands of feet higher. A. A. Miller (1946) notes that the lowest temperature on record in the USA ($-65°F$ [$-54°C$]) was reached at Miles City, Montana, lying in a deep hollow in the Great Plains, while Pike's Peak, which is 11 000 feet higher, has never recorded a temperature below $-40°F$ [$-40°C$].

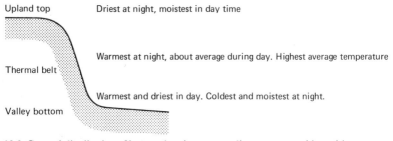

Upland top — Driest at night, moistest in day time

Thermal belt — Warmest at night, about average during day. Highest average temperature

Valley bottom — Warmest and driest in day. Coldest and moistest at night.

12.2 General distribution of heat and moisture according to topographic position

These movements of air are reflected in similar movements in cloudiness and rainfall. In the valley bottoms, both are at a maximum at night and in the winter, but at a minimum in the day time and in summer. This is because the greater convectional rise when the sun's heat is strongest carries the moisture upwards to a higher altitude before dew point is reached and clouds and rain are formed.

Relationships of minor terrain features to microclimate

Another topographic factor important to climate is aspect. Maximum insolation occurs on any surface normal to the sun's rays, i.e. with a southerly aspect in the northern hemisphere. This means that for any given latitude, and neglecting broad climatic effects such as seasonal and daily variations in cloud cover, the aspect giving the maximum radiation receipt is that of a south-facing surface inclined so that it will receive the sun's rays most nearly vertically for the maximum duration during the year. It can probably be assumed that 'average' conditions during the year occur at the equinoxes. This would indicate that, for greatest insolation over the year, the equatorward inclination of the land surface should be approximately equivalent to the latitude, i.e. 52° in London, 56° in Edinburgh, etc.

Exposure can be defined generally as accessibility to climatic effects, notably wind. Some analogy may be seen between the exposure of terrain, on which almost no research has been done, and that of buildings, whose heat transmission through walls and resulting insulation and heating requirements have been the subject of study (e.g. Dufton, 1940–41). Certain empirical values have been worked out for Britain based on a system of 'penalty points' by which sites are downgraded according to the unfavourability of their orientation and the degree of severity of their exposure to wind.

The values are assessed according to the following table:

TABLE 12.1. Penalty points for the unfavourability of the orientation and exposure of buildings in Great Britain

Orientation:	S	0
	W, SW, or SE	2
	Other	4
Exposure:	Sheltered	0
	Normal	2
	Severe (except N, NE, or E)	4
	Severe (N, NE, or E)	6

As the orientation and exposure values are additive, the possible range is from 0, for a sheltered site facing S, to 10, for a severely exposed site facing N, NE, or E. This scheme would not, of course, be suitable without modification outside Great Britain, and would be irrelevant outside the North Temperate Zone.

Aspect and exposure are generally reflected in land uses and land values. In mountainous parts of Switzerland there is the well-known contrast between the south-facing *Sonnenseite* and north-facing *Schattenseite* of valleys, the former having higher snow lines, a longer growing season, and more favourable conditions for human settlement. In the wine growing districts of France and Germany, vineyards are located on south-facing slopes to avoid the exposure of the plateau tops and the frost pockets of the valley bottoms. In temperate countries generally, the value of residential property varies appreciably with aspect.

The effects of meso- and microrelief are similar to those of macrorelief, but are more ephemeral and variable, especially when the topography is broken or complex. Vegetation, the works of man, such as buildings and earthworks, and even such micro-features as plough ridges are significant, especially to ground temperatures.

As a general world average, about 35 per cent of the incoming solar energy is reflected directly back into space from the atmosphere and the earth's surface and lost. The atmosphere absorbs a further 19 per cent. Only the remaining 46 per cent is retained by the ground. Different types of ground surface reflect different proportions of this back into the atmosphere. This reflected portion is known as the *albedo* and is expressed as a percentage of the total radiation striking the ground, according to the formula:

$$R + D + A = 100\%,$$

where:

R = albedo (per cent reflected)
D = radiation penetrating to the ground (per cent)
A = radiation absorbed by plants (per cent)

Table 12.2 shows the measured albedos for certain types of ground surface.

TABLE 12.2. Total surface effect of different types of material with diffuse reflectance (after Geiger, 1965)

	ALBEDO VALUES
Snow	20–95 (higher values when fresh and clean)
Light sand dunes	30–60
Glacier ice	20–46 (higher values when clean)
Sandy soil	15–40
Meadows and fields	12–30
Densely built-up areas	15–25
Woods	5–20
Dark cultivated soil	7–10
Water surfaces	3–10

Neglecting the effect of vegetation, the type of surface material affects the regime and depth of penetration of surface heating. Observations in Finland and Japan, quoted by Geiger (1965), showed that the temperatures of exposed ground surfaces were at a maximum in conductive rocks such as granite, with decreasing values in soils in the order sands – loams – clays – bog, although sands may give higher daytime temperatures than granite. The depth at which daily temperature variations exceed $0.1°C$ decreased in the same order from over 60 cm in granite to 40 cm in bog.

Both high bulk density and high moisture content in soils narrow the range of surface temperature variations and delay the arrival of maxima and minima. The effect of snow is less cold than might be expected because, although its mass and surface are at low temperatures and have high albedos, it allows underlying soil to retain a relatively high temperature due to its blanketing effect, while at the same time, being a better conductor than water, it transmits solar energy to the ground.

The darker the colour of the ground surface, the lower is the albedo. This means that the heat uptake of soils can be increased by covering the surface with darker coloured materials, such as mulches or stones.

Vegetation interposes a barrier between the atmosphere and the ground surface, which intercepts a proportion of the rainfall and diffuses both inward and outward radiation over the whole zone of plant height. Even in a meadow, the surface area of leaves is 20 to 40 times as great as that of the ground on which they grow, making natural transpiration greatly in excess of evaporation. This is so important in forest that the vegetation can be said to regulate the whole water regime, reducing the soil moisture content in direct relation to the volume of the vegetation, so that it is drier under trees than under low vegetation. Wind strength is also greatly reduced and temperatures are milder under trees than in the open. On the whole, vegetation has an 'oceanic' effect on the climate of the subjacent terrain in the sense of narrowing temperature ranges, delaying extremes, and increasing the amount of moisture in air and soil. It also has such minor effects as causing snow to melt more quickly immediately beside dark-coloured heat-absorbent tissues, such as tree trunks, than elsewhere.

In sum, a climatic classification of terrain must begin with a broad subdivision of the landscape into zones of the Köppen – Thornwaite type. These can be subdivided at intermediate scale on the basis of altitude and general relief configuration, and at small scale on the basis of aspect, exposure, and surface character to the degree to which these reflect economically significant differences in isolation, temperature, and rainfall regime. Fig. 12.3 gives a highly schematic impression of how a landscape might be classified on climatic grounds.

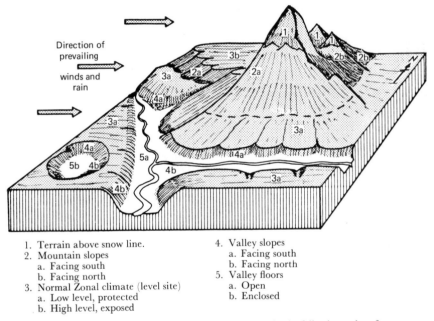

Direction of prevailing winds and rain

1. Terrain above snow line.
2. Mountain slopes
   a. Facing south
   b. Facing north
3. Normal Zonal climate (level site)
   a. Low level, protected
   b. High level, exposed
4. Valley slopes
   a. Facing south
   b. Facing north
5. Valley floors
   a. Open
   b. Enclosed

Considering only climatic favorability, the sites might be in the following order of economic value: 4a, 3a, 2a, 5a, 3b, 4b, 5b, 1, 2b.

12.3 Proposed broad climatological classification of a schematic landform assemblage in the North Temperate Zone

13
Terrain factors in hydrology

The major hydrological unit: the river catchment

The water in any area is contained either in surface bodies such as rivers or lakes, in the superjacent air, or underground. The science of hydrology is concerned with the quantification of water in these three states and the exchange between them, while applied hydrology is concerned more specifically with the assessment of water resources and the prediction of forthcoming events, notably floods and droughts.

The primary unit of terrain applicable to hydrology is the river catchment or basin, and this is generally adopted as both the scientific and administrative unit. It forms the basis for the measurement of all the critical parameters: total precipitation, evaporation from soil and open water surfaces, transpiration from plants, surface runoff and storage of water, infiltration into the soil and to the ground water, and underground flow.

Water enters a catchment in the form of rain or snow. Its quantity can be assessed by the Thiessen, the isohyetal, or the representative basin methods, described by More (1967). Briefly, the Thiessen method assesses overall rainfall in a catchment by adding up the areas geometrically nearest to climatic stations whose rainfall values are taken as representative. The isohyetal method deduces the same value by summation of the areas and rainfall amounts between the regional isohyets, while the representative sample method extrapolates values from small, carefully measured sub-catchments chosen as typical of the main one.

Evaporation can be measured from a free water surface with various types of evaporating pans, but the errors and difficulties involved in applying these values to land surfaces, especially where vegetation is present, have led to the derivation of empirical formulae based on the concept of energy balance. Penman's (1963) formula equated energy input and output, using measurements of air temperature, wind speed, saturation deficit, and day length. Somewhat similar formulae applicable to different regions have been derived by Blaney and Criddle (1950), Thornthwaite (1948 and 1954), and Olivier (1961). None of these methods incorporate a factor for terrain type.

Terrain factors determine the rapidity of runoff and the character of flow maxima and minima in streams after storms. For instance, a topography of steeply sloping, smooth surfaced, impermeable materials with few pondage pockets and little vegetation gives very rapid runoff and high sudden flow maxima, while the opposite conditions will give a long continuous flow with low and delayed maxima.

Actual measurements of river flows depend on channel dimensions and condition, but can be calculated approximately from measurements of cross-sectional area and current speed. On the whole, channels become wider and shallower relative to their depth and adjusted to larger flows as they progress downstream. Bankfull discharges increase in the same direction in proportion to the square of the width of the channel or of the length of individual meanders and in proportion to the 0.75 power of the total drainage area focused at the point in question.

Empirical calculations to predict river flows can be made by hydro-meteorological or statistical methods, after making allowance for infiltration and evapotranspiration. Hydrometeorological analysis is based on an estimate of the highest and lowest values which can reason-ably be expected from all the meteorological factors involved in causing the extreme conditions and the assessment of their likely effect when acting in concert. Statistical methods start from the assumption that recurrence intervals of extreme events bear a consistent relationship to their magnitudes. A recurrence interval (e.g. fifty years) is chosen in accordance with given hydrological requirements. The risk of relying on these assumptions in practise necessitates the use of the first method in combination with the second to ensure a margin of safety for hydro-logical purposes (More, 1967).

Because of the complexity of measuring these indices of riverflow, approximations have been sought which give a satisfactory integration of the hydrological characteristics of river basins. An important example of this is Sherman's (1932) 'unit hydrograph' concept, which postulates that the most important hydrological characteristics of any basin can be seen from the direct runoff hydrograph resulting from 1 inch (25 mm) of rainfall evenly distributed over twenty-four hours. This is produced by drawing a graph of the total stream flow at a chosen point as it changes with time after such a storm, from which the normal base flow caused by groundwater is substracted.

The subdivision of major catchments

River catchments can be divided into subcatchments of decreasing size, with those of the smallest tributaries as the ultimate basic units. Horton, in 1945, gave this a quantitative basis by classifying channels and basins into a hierarchy of *stream orders*. His scheme was simplified by Strahler (in Chow, 1964), and has stimulated much research, summarized by Haggett and Chorley (1969). Because of the tendency of heavy storms to be local in occurrence, stream flows tend to be most variable and flood discharges at a maximum per unit area in the smallest basins.

Stream order 1 includes 'fingertip' streams receiving no tributary. Stream order 2 includes those which are formed by the junction of two first-order streams and only receive first-order tributaries. Stream order 3 includes those which are formed by the junction of two second-order streams and only receive first and second-order tributaries, and so on. Fig. 13.1 makes this classification clear.

Both the number and the overall length of the streams within a drainage basin decrease geometrically with stream order. These facts are illustrated mathematically by two ratios: the *bifurcation ratio* and the *stream length ratio*.

The bifurcation ratio for any consecutive pair of orders is assessed as the total number of streams of the lower order divided by the total

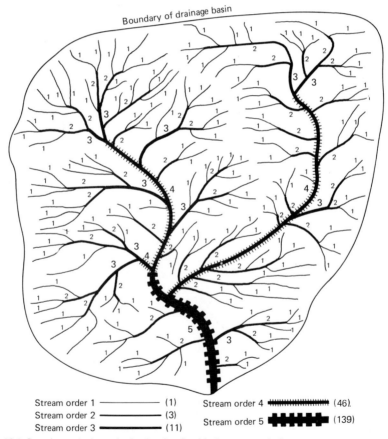

Stream order 1 ——————— (1)	Stream order 4 ┉┉┉┉┉┉ (46)
Stream order 2 ——————— (3)	Stream order 5 ▪▪▪▪▪▪ (139)
Stream order 3 ——————— (11)	

13.1 Imaginary drainage basin showing Strahler's stream ordering system

number in the higher order. It is the most frequently used of the two, partly because of its greater ease of measurement. There are several other ways of assessing it, which are summarised by Haggett and Chorley (1969). The most commonly used, however, is the method which allows a single figure to be calculated for a whole drainage basin. This is derived by drawing a graph of the relationship between stream order along the x axis, and the logarithm of the number of streams along the y axis. The slope of the 'best fit' line joining the resulting points is the *regression coefficient* and the bifurcation ratio is the antilogarithm (to the base 10) of this value. Fig. 13.2 makes these relationships clear. Alternatively, the bifurcation ratio can be derived simply by averaging the ratios from the numbers of streams in each consecutive pair of orders. For instance, for a basin with 3 orders, the formula would be as follows:

Bifurcation ratio

$$= \left[\frac{\text{number of streams in order 1}}{\text{number of streams in order 2}} + \frac{\text{number of streams in order 2}}{\text{number of streams in order 3}} \right] \div 2$$

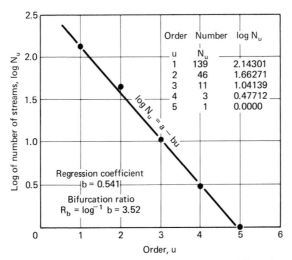

13.2 Regression of a number of stream segments on order, giving bifurcation ratio of Fig. 13.1 (Source: P. Haggett and R. J. Chorley, 1969, p. 18)

Where rock and structure are simple and do not form narrow elongated basins, bifurcation ratios seem to be stable and generally range between about 3 and 5. The pattern of runoff in basins seems to reflect this ratio.

The stream length ratio is assessed as the total length of streams of the lower order divided by the total length of those in the higher. Values for this ratio depend mainly on drainage density and stream entrance angles and increase somewhat with increasing order.

A similar measure is the *texture ratio*, suggested by Smith (1950). This is the number of crenulations in the basin contour having the maximum number of crenulations, divided by the perimeter of the basin. This has been shown to be related to drainage density by Strahler (1957).

Each basin has an identifying *basin order*, defined numerically by the highest order stream it contains. There is a general relationship between the order of a basin, the discharge of its stream, and the square root of its total drainage area. The *drainage density* expresses the fineness of erosional texture of the landscape and is defined as the average length of stream channel per unit area of basin. It is derived by dividing the total length of the stream channels in a basin by its area and is generally expressed as miles per square mile. Many measurements of this factor have been made in different parts of USA by different workers, and have been summarised by Haggett and Chorley (1969). They range all the way from 3 to 1300, with lowest values generally in areas of hard, permeable rocks under deciduous forest and highest in badland-type areas of soft impermeable clays and shales under semi-arid conditions.

Subdivision of Order 1 catchments

The ultimate hydrological classification of terrain subdivides Order 1 catchments in terms of the interrelated factors of gradient, infiltration capacity, and surface roughness. These factors determine the proportion of water entering the soil to feed the ground water, running off into the drainage channels, or lost by evapotranspiration, and permit the terrain of the catchment to be subdivided into its hydrologically significant parts.

The water falling on a basin which neither infiltrates into the ground nor returns to the air by evapotranspiration begins as unconfined overland flow. This is the normal state everywhere away from stream courses and other water bodies. Horton (1945) has noted that around the edge of every catchment there is a zone where all flow is of this nature before it concentrates into channels. He designates this as x_c and defines it as the critical length of overland flow from the hillcrest required to produce enough runoff to start erosion and therefore to change from overland to confined flow. This distance depends on gradient, runoff intensity, soil infiltration capacity, and the roughness and resistance to sheet erosion of the surface. It can be expressed by the formula:

$$x_c = \frac{65}{q_s n} \times \left(\frac{Ri}{f(s)} \right)^{5/3}$$

where,

x_c = the width of no erosion, defined as the horizontal length of the flow path from a point on the divide to the approximate ortho-gonal point on the adjacent stream channel (in feet).

q_s = the surface runoff intensity (in inches per hour). This usually ranges from 0·5 to 2, 1 being an approximate average.

$n =$ the surface roughness factor, as in the Manning formula. This varies with type, height and density of vegetation, roughness of land, and in flow channels, with their size, cross section, alignment, and hydraulic radius i.e. generally the conditions which increase turbulence and retard flow. It is assessed as follows (Cowan, 1956): first, a basic n value is selected for the material under consideration. 0·02 is a useful figure for earth, 0·025 for rock, 0·024 for fine gravel, and 0·028 for coarse gravel. To these basic values are added numerical assessments in direct proportion to the resistance to flow posed by four further characteristics: surface irregularity: 0–0·02; obstructions (such as debris, stumps, logs, roots, etc.): 0–0·06; vegetation: 0–0·05; and degree of meandering 0–0·3. The results of all assessments are then added together to give n. In general, in smooth materials it lies between 0·012 and 0·024; in earth without vegetation it increases from 0·025 to 0·06 with increasing coarseness of texture, and in vegetated channels it increases from 0·04–0·2 with increasing size of vegetation (Schwab *et al.*, 1966).

$Ri =$ the initial surface resistance to sheet erosion (in lb per square foot). This varies from about 0·1 for high resistance, i.e. where there is high infiltration capacity and a dense vegetation cover, to 0·5 for low resistance.

$f(s) =$ the effect of slope on x_c. This is given by the formula:

$$f(s) = \frac{\sin a}{\tan^{0\cdot3} a} \qquad \text{where } a \text{ is the angle of slope.}$$

In first-order drainage basins, the x_c factor seems generally to be approximately equal to one half the reciprocal of the drainage density.

The land surface determines the amount of water which can be stored in lakes, basins, and reservoirs by controlling their overall area, depth, and the permeability of their beds. The larger and shallower the basin, the greater the area over which evaporation and downward percolation can take place and hence the greater the rate of water loss.

Terrain conditions also govern the quantities and rates of recharge to, transmission and storage within, and loss from, the ground water reservoir. Recharge is that fraction of the rainfall which is neither lost through evapotranspiration nor diverted by runoff. Its relative amount is directly related to surface roughness and permeability and inversely to gradient. Permeability is most accurately defined as the 'k' (hydraulic conductivity) factor in Darcy's flow law and is generally a function of soil particle size. Roughness is derived from the 'Manning coefficient' discussed below. The velocity of ground water transmission is defined as the product of the hydraulic conductivity and the piezometric head loss divided by the distance over which the movement is occurring. The quantity of this transmission per unit time over a given cross-sectional area can be derived by multiplying the velocity by the area. Fig. 13.3 makes these relations clear.

111

The effective water storage of a subsoil aquifer is given by the volume of its non-capillary pores. This can either be deduced by multiplying the total volume of the aquifer, determined by geological or geophysical methods, by a value for pore space derived from consideration of its particle size composition or, alternatively, by measuring the output of sample wells.

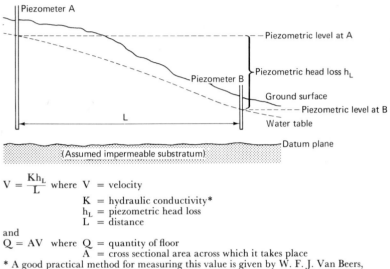

$$V = \frac{Kh_L}{L} \quad \text{where} \quad V = \text{velocity}$$

$\qquad\qquad\qquad K = \text{hydraulic conductivity*}$
$\qquad\qquad\qquad h_L = \text{piezometric head loss}$
$\qquad\qquad\qquad L = \text{distance}$

and

$$Q = AV \quad \text{where} \quad Q = \text{quantity of floor}$$

$\qquad\qquad\qquad A = \text{cross sectional area across which it takes place}$

* A good practical method for measuring this value is given by W. F. J. Van Beers, 1958.

13.3 Diagram and formulae to illustrate underground water movement

Groundwater emerges at the surface in the form of springs, which occur most commonly at the outcrop of impermeable underlying permeable rocks on slopes, or in the form of evaporation from shallow groundwater tables through soil capillaries and transpiration by plants, especially phreatophytes, in hot arid climates.

The second zone in the basin begins with the commencement of rills at accidental points of low resistance on the hillside below x_c. The deepest and widest rills form where the net length of slope below x_c is greatest and grow by micropiracy of smaller parallel rills. As they become established, these rills become the foci of secondary drainage channels by cross-grading. This process is illustrated diagrammatically on Fig. 13.4.

The third zone is the stream and river channels themselves, which have the characteristic forms described in chapter 3.

Infiltration capacity increases with surface roughness and permeability and decreases with gradient. It therefore tends to be greatest in the relatively rough, course-textured, permeable, and more gently sloping beds of channels than on the intervening terrain.

Surface roughness, previously defined as Manning's coefficient n, is related to the speed of flow by the Manning formula:

$$v = \frac{1\cdot49}{n} R^{2/3} s^{1/2}$$

where,

v = velocity, in feet per second.

n = surface roughness.

R = the cross-sectional area of flow divided by the wetted perimeter, in feet.

s = the hydraulic gradient.

Permeability is closely related to granulometry and soil structure, both of which can be measured by a number of field and laboratory methods. Some of these are outlined in chapters 16 and 17. The relevance of soil infiltration rates to hydrology is shown by their proved relationship both to durations and to times-to-peak of river discharges after rainstorms in north eastern England (Edmonds, Painter and Ashley, 1970).

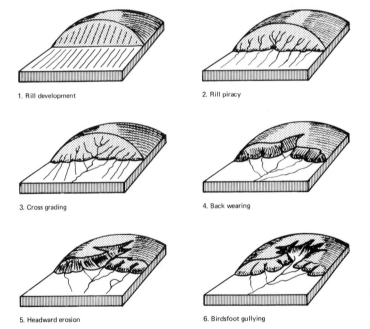

1. Rill development

2. Rill piracy

3. Cross grading

4. Back wearing

5. Headward erosion

6. Birdsfoot gullying

13.4 Sketch to illustrate the development of a drainage channel on an original cut slope in uniform materials (Adapted from Horton, 1945)

To summarise, a hydrological classification of terrain must treat drainage catchments as the basic units, beginning with those of stream order 1. Each such catchment must then be subdivided in terms of its slopes, surface roughness, infiltration capacity and underground storage capacity. When such a subdivision is combined with an assessment of the local balance between rainfall and evapotranspiration an accurate idea of the water holding and transmitting characteristics of the catchment is obtained.

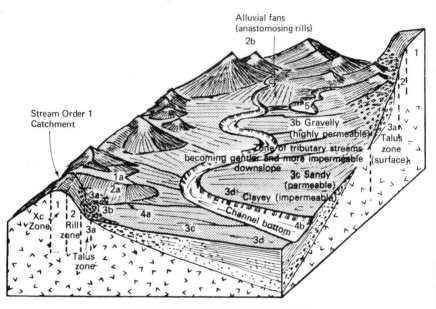

HYDROLOGICAL TERRAIN ZONES
1. Xc Zone at catchment edge: underground percolation, no rills.
2. Rill zone: steep.
3. Lower drainage slope with meandering channels:
   a. Talus
   b. Upper, highly permeable (gravelly)
   c. Middle, permeable (sandy)
   d. Lower, impermeable (clayey)
4. Stream channels:
   a. Stream order 1
   b. Higher stream order
5. Lake or reservoir basins.

13.5 A schematic classification of terrain according to hydrological factors in a small catchment and the larger one to which it is tributary

114

14
Landscape resource analysis for landscape and recreational planning

It is sometimes forgotten that landscape has a definite economic value for uses other than agriculture, buildings or public works. Its aesthetic and recreational value is being more and more widely recognised. The Bernese Oberland is nothing but a waste of rock and snow while most of the Costa Brava is unsuited to any sort of agriculture, but both attract millions of pounds of tourist money annually. These are extreme examples, but even within fairly uniform landscapes it is possible to discriminate areas of greater and less aesthetic value. The North Downs and Chilterns provide some of the best recreational and residential land within the immediate vicinity of London, while the locations of spas such as Tunbridge Wells and resorts such as Brighton show the economic attractions of hilly or coastal sites.

A number of methods have been evolved for putting the assessment of landscape aesthetics on a more quantitative basis. The first step is the subdivision of the area under consideration into planning regions. These are almost invariably political or administrative units or the drainage catchments of major rivers. Inventories are then made of the environmental resources and of the demand for them. As Steinitz (1970) points out, these are then linked into 'simulation models' which can help forecast future trends. Understanding these trends enables the planner to predict the changes likely to take place in the landscape, in the characteristics of public demand for it, and in such related factors as the transport networks which serve it.

Urban and rural landscape planning are sufficiently distinct operations to be considered separately. In towns, the visual scale is smaller and more confined, so that aesthetic problems are mainly architectural, competition for the use of land is more severe, and aerial pollution is of greater relative importance. In rural areas, views are wider, recreational use less intensive, and conservation is more concerned with the protection of the landscape against erosion and soil and water pollution.

Urban land includes all built-up areas: residential, central, commercial, industrial, and transport, all of which generally require flat, well drained sites with bedrock deep enough below the surface to avoid constructional and drainage problems. Residential property differs in preferring somewhat steeper gradients and in being able to accommodate itself to smaller and less compact blocks of land. Industrial plants and regional and commercial centres require larger expanses of flat land. Kiefer (1967) has suggested a scheme by which land use suitability ratings can be combined with terrain maps to give 'relative land use suitability' maps to help guide urban developments. The terrain maps are based on a combination of parent material, slope class, soil class (according to the USCS system), and drainage condition. Each of these attributes is separately mapped and then the results are overlaid. The highest topographic rating is given to slopes of 2–5° as they avoid the more serious drainage problems of flatter lands and the engineering problems of building on steeper slopes. Evaluation of soil class is based on its strength, stability, and permeability (10–60 cm per day being optimal), and on there being an adequate depth to water table. Depth to bedrock should exceed 210 cm, and an especially low rating is given if it is shallower than 90 cm. Combined ratings are based on those of the most limiting characteristic.

Land used for transport purposes, such as roads, railways, canals, and airports, demands special treatment because of their peculiar spatial demands and the specific two-way interaction with landscape imposed by moving traffic. While on the one hand, highways provide the means for bringing most people into contact with the countryside, on the other hand, they bring danger, noise, and air pollution to their surroundings. Land near highways is also peculiar in that it is usually viewed while the observer is in motion. Economic considerations alone are an unsafe guide to choice of routes, and it is now recognised that, at least for developed areas, the overriding principle must be to provide maximum social benefit at least social cost (McHarg, 1967). Road alignments should, therefore, generally avoid traversing land with either important physiographic obstructions or high social values. The former depends on slope lengths and gradients. The latter is harder to quantify, but is directly related to the level of land and building prices and of agricultural, historical, hydrological, wildlife, and scenic considerations.

Some attempts have been made to assess landscape quality from the point of view of the traveller. One example is the semiquantitative scheme for judging the proposed route of the Durham Motorway from Darlington to Chester-le-Street described by J. B. Clouston (Landscape Research Group, 1967). A composite map (part of which is reproduced as Fig. 14.1), was produced, which gave a simple landscape assessment of the *visual corridor* along which a passing driver would travel, modified by the *vision frequency*, i.e. the number of times he would see any part of this corridor. Its width was in part determined by the frequency of poor visibility in the area. The method allowed an optimal route to be chosen both from the point of view of the motorist and nearby land users.

Rural or undeveloped terrain fulfils a considerable variety of practical, aesthetic, and recreational needs, which can be divided into three main groups. First, and most intensive, is the land which is used for team

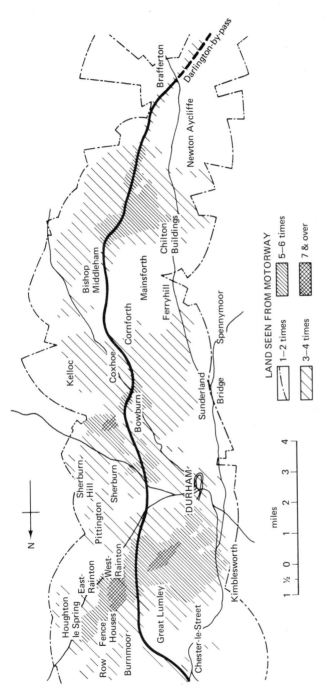

14.1 The use of visual criteria in road planning. In a study for the Durham Motorway land was classified in accordance with its degree of visibility from the proposed route line with a *visual corridor* and areas of *vision frequency* (Source: J. B. Clouston, 1967, opp. p. 15)

LAND SEEN FROM MOTORWAY

1–2 times
3–4 times
5–6 times
7 & over

sports and children's group play. Individual areas may be small but are frequent where population is dense. The land is flat and level. In humid climates it needs to be well drained, so that it can be used soon after heavy rain, while in dry climates it must be neither too hard and stony nor too soft and dusty.

Secondly, there is rural land which is suited to scientific and educational purposes. This includes nature reserves, botanical and zoological collections, arboreta, field studies centres, and research and demonstration centres of various kinds. The dominant concern is not for aesthetics, but for the well-being of the plant and animal species involved. Sites are most suitable which represent one or other of two extremes: in being either representative of an area or else very exceptional. An agricultural research centre, for instance, needs to occupy a site typical of the area it is designed to serve, while a nature reserve seeks to preserve some unusual conditions of rock, vegetation, or habitat.

Thirdly, there is terrain suited to more individual and wider ranging forms of sport and recreation. This includes both local and national parks. Local parks include public gardens as well as cemeteries and memorial grounds devoted generally to leisurely exercise and calm enjoyment. National parks are larger and wilder, providing special terrain conditions in a rural setting. They have footpaths for running, rambling, and picnicking, bridle paths, golf courses, and open water with beaches for swimming, boating, sailing, and fishing accessible to large population centres. These are the only classes of recreational land whose choice and management depend primarily on the visual and aesthetic qualities of landscape.

The general quantification of landscape in terms of enjoyment value depends on a host of different considerations, both of the environment itself and in the eye of the beholder. Certain general principles can, however, be suggested on commonsense grounds. Landscape, to be attractive, must have *contrast*. This can be between light and dark, vertical and horizontal, straight lines and curves, foreground and background, land with water and sky, and many others. There should generally be an absence of stiff, formal angles or of undue repetition of the same elements. Bold, abrupt foregrounds have a softening effect on distance. Curves should generally be gentle and wide, except when leading up to notable features.

Colours should be in harmonious combination. Two general principles seem to be valid. First, as pointed out by Dennis (1835), if one arranges the colours of the spectrum in order from 1 to 9 thus:

1. White	4. Yellow	7. Indigo
2. Red	5. Green	8. Violet
3. Orange	6. Blue	9. Black

it will generally be found that neighbouring colours blend easily, that the most harmonious combinations are those between alternate colours, i.e. 1 with 3, 4 with 6 etc., but that contrasts are usually strong enough to be unattractive when more than one intervening colour is jumped. A second principle should be considered in conjunction with this as showing where strong contrasts can be most attractive. It teaches that each of the three primary colours red, blue, and yellow will go well with a combination of the other two, i.e. red with green, purple with yellow, and so on.

Physiographic features which are generally beneficial in the landscape are those which give attractive contrasts in texture, orientation, and shape, as well as colour. Hills and mountains, especially where steep or snow capped, give dramatic quality to a scene. Open water is always advantageous, particularly when it is in motion and includes pools, waterfalls, boulders, or other eye-catching features. Coasts are generally attractive and are improved from the recreational point of view by the presence of sandy beaches and rock pools. Woodland adds charm to all scenes, and is valued by the tourist when open enough to be generally penetrable to the rider or walker, and to include glades.

The physical comfort of the viewer must also be considered. Recreational land should be dry and well drained so that it does not form puddles or mud which make walking and sitting unpleasant. This is one reason why the freely draining gravelly Bagshot beds make such attractive recreational and residential property in the London area. Sites should also be placed favourably in relation to sun, wind, and access routes. Under European conditions this means sheltered southerly orientations with convenient but inconspicuous roads.

While larger natural features are unchanging and must be exploited as they are, smaller ones may sometimes be modified. Paths should not be too sharply curved or too steep. In a jointed rock, it is sometimes possible to fill hollows with decorative plants. Man-made modifications of terrain such as pleasaunces, ha-has, banks, rockeries and grottoes are often valuable.

A number of methods have been suggested to place the assessment of the aesthetic qualities of landscape on a more quantitative and less subjective basis. The simplest approach is to classify a given landscape in relation to different types of recreational land uses and to draw a map of the results. An example of this general method is Lewis's study of Wisconsin in 1964 to help evaluate a proposed recreational 'heritage trail' as part of a network of 'tourist trails' in that state; a part of it is shown in Fig. 14.2. It identifies the four main types of area attractive to the tourist: those with 'significant' water, topography, wetlands, or landscape personality. The major roads and trails of the state are superimposed on this base.

Similar schemes have been devised by a number of county councils and other bodies in Britain for the evaluation of landscape. The British Trust for ornithology, for instance, classified the landscape in terms of its suitability as a habitat for different birds. The Hampshire Planning Office assessed the aesthetic quality of areas on the basis of a number of empirical criteria observed at a grid of sites one half mile to one mile apart. Good marks were given for favourable factors such as the amount of vertical relief, and bad ones for 'detractors' such as neglected land, rubbish tips, battery chicken houses, etc. A similar scale of assessment was evolved by East Sussex Planning Office. Land was viewed at a

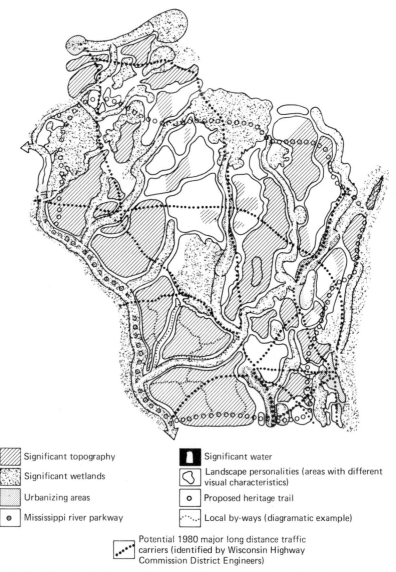

Significant topography

Significant wetlands

Urbanizing areas

Mississippi river parkway

Significant water

Landscape personalities (areas with different visual characteristics)

Proposed heritage trail

Local by-ways (diagramatic example)

Potential 1980 major long distance traffic carriers (identified by Wisconsin Highway Commission District Engineers)

14.2 The Wisconsin 'heritage trails' proposal (Source: P. H. Lewis, 1964, p. 102)

network of points and graded from 'disfigured' to 'spectacular', specifically including a weighting for weather conditions. These schemes are little more than exploratory, however, and there is still a long way to go before a common quantitative standard for grading landscape is agreed even for Britain.

Tandy (1967) has attempted to refine this grid approach both in concept and in scale by introducing his *isovist* method. This begins with the subdivision of landscape into small areas centred on points called *focal centres* at selected heights above ground and placed so that each commands at 360° circle of vision. The area visible from each point is called the *visual zone* and its enclosing line is called the *isovist line* in rough analogy with isobars, isotherms, etc. When isovist lines around several focal points are plotted on a map, they tend to overlap considerably. The overlapping areas are simplified by inspection to obtain single compromise isovist lines enclosing distinct zones with clearly defined identities. These zones vary in character but can be classified into types to which names are given, e.g. *extroverted, introverted, linear, outward looking* (one-direction), open-ended, etc., as illustrated on Fig. 14.3. When two zones are separated by a ridge or other strongly limiting feature, the boundary is considered a *visual watershed*. The overall appraisal of an area is based on an evaluation of the visual quality of each constituent zone followed by a synthesis of them all.

In certain districts a single feature so dominates the landscape that special means must be found for assessing its impact. Power stations are an example of such features. They are often large, ugly buildings which detract from the beauty of the country areas in which they are situated. Murray (1967) has devised a method for analysing their general appearance in relation to the numbers and types of observers who view them, the circumstances of viewing such as weather, time, and season, and the activities being carried out by observers.

Evaluation is based on five mathematically determined factors for each potential viewpoint:

1. *The silhouette factor*: the visible area of silhouette from a given viewpoint according to an arbitrarily chosen scale. This can also be expressed visually by building a scale model or by superimposing a scaled silhouette of the proposed station on to a lantern slide of the view.

2. The *distance coefficient*: the ratio between 25° (the 'normal effective vision cone') and the greatest angle subtended by the silhouette.

3. The *visibility coefficient*: the proportion of the days in the year in which the visibility is inadequate to see the station.

4. The *displacement coefficient*: the angular distance of the station from an observer's preferred line of sight. When the viewpoint is one used by someone driving a car, the displacement coefficient must be modified to take account of the decreasing cone of vision which results from increasing driving speed. This modification is known as the *speed coefficient*.

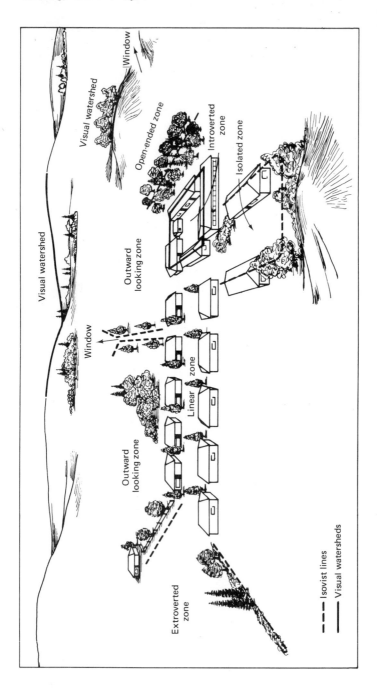

14.3 Sketched example of an *isovist* summary of a landscape, according to the method suggested by C. R. V. Tandy, 1967.

The author does not specify exactly how these dimensionless coefficients are to be applied. Their value lies in their comprehensiveness in quantifying visual aspects of power station rating in terms of the locations and conditions of viewing. They make it possible to compare sites numerically in relation to a given viewpoint or to compare viewpoints numerically in relation to a given site.

Jacobs and Way (1969) have developed a scheme of landscape evaluation based on the concepts of 'visual transparency', and 'visual complexity'. Visual transparency is defined by the degree of vegetation density and the amount of topographic closure. Visual complexity is defined by the amount and the clarity of visual information which the viewer must sort and evaluate. These parameters are assessed on the ground using arbitrarily determined numerical scales. The significance of the indices lies in the fact that building developments and other man-made changes in the environment will be least obtrusive and hence generally least undesirable in landscapes which are most opaque and complex.

Part Three

Methods in Terrain Evaluation

15
Preliminary assessment of land

Exploitation of all existing sources of data

When an area is to be evaluated, the first step is to undertake a systematic review of all the available information about it in order to arrive at a preliminary assessment. Such information is usually in the form of books, articles, maps, unpublished reports, and aerial photographs.

Books and articles are relatively easy to obtain. For comprehensive coverage it is usually necessary to make use of the libraries in the country in which the survey area lies. The key background is that on geology and soils derived from such bibliographies as the US Geological Survey *Bibliography of North American Geology* and the Geological Society of America *Bibliography and Index of Geology exclusive of North America*, both published annually, and from the monthly *Geomorphological Abstracts* in Britain.

Topographic maps exist for all the land surface of the earth, but the best scales available vary widely. While for developed countries much coverage exists at scales larger than 1:50 000, 1:100 000 is more common in developing areas, and most of their more remote parts are mapped at only 1:250 000. Geological and other thematic maps are rarer and at smaller scales. Although most of Britain, for instance, is covered by geological mapping at 1:63 360, some parts are only at 1:250 000. Most of Africa and Asia, however, is represented at only 1:2M or smaller. Soil and other thematic maps are still more infrequent and at even smaller scales.

Detail and scale of survey

This depends on the purpose for which a survey is being carried out, the time available for its completion, and the amount and nature of prior knowledge available. There are however, three clearly recognisable stages (Robertson *et al.*, 1968).

1. *First stage regional surveys.* In countries where knowledge of natural resources is insufficient to indicate development possibilities, the initial requirement is for rapid general surveys to show the range of such possibilities and where more detailed investigations will be most

rewarding in accordance with an overall view of national development priorities. Such surveys are generally based on the land systems approach and the interpretation of aerial photographs at scales of 1:40 000 or smaller.

2. *Second stage surveys* analyse the development possibilities in a project area or group of areas which have been identified in the first stage. This involves the acquisition of sufficient information to indicate whether specific development projects are feasible, both from the physical point of view and from the general cost/benefit aspect. This means that the scale of survey has to be enlarged to give fairly detailed information on the various resources involved. This involves the inclusion of soil surveys, the quantification of vegetation resources, the identification of water storage sites, the measurement of groundwater supplies and quality, and an assessment of present land uses.

Work at this stage will generally be based on maps and aerial photographs at scales of 1:25 000 or larger.

3. *Third stage, or detailed, surveys* are analyses of an area to formulate a specific project plan, and to show in detail the cost/benefit ratios of the activities considered. Studies are made of the crop, mineral, or manufactured item to be produced and its management in terms of likely costs and yields. Scale of observation need not be much, if any, larger than in second stage surveys but there will be inevitably a heavier emphasis on detailed economic assessment.

The use of remote sensing

Remote sensing is today one of the basic preliminaries to land assessment. It can be literally defined as any perception and recording of phenomena by devices not in contact with them. Specifically, remote sensing of the terrestrial environment is the identification and analysis of phenomena on the surface of the earth using devices carried in aircraft or spacecraft. These devices depend on the sampling of radiant energy within the electromagnetic spectrum. Within this spectrum, energy moves at a wide range of wavelengths and frequencies. These are summarised in Table 15.1.

The principal source of energy is the sun and most sensing devices are passive in that they depend on solar radiation which has been received at or near the earth's surface and radiated back into the atmosphere. On the other hand, some sensors, such as radar, are active in that they generate their own energy and record it as it returns to source.

Another group of sensors are those used by geophysicists to penetrate deeper into the earth's crust than is normally needed in terrain evaluation. These include magnetometers to record anomalies in the earth's magnetic field, gravity meters to record variation in its gravitational field and hence the type of rock mantle, and scintillation counters to indicate radioactive areas. All these tools can be used either on the

127

ground or from aircraft.

Passive direct systems of remote sensing from aircraft are photographic systems which employ the visible and near-IR wavebands of the electromagnetic spectrum, usually reproduced as photographic prints or transparent diapositives. Such photographs are still by far the most important remote sensing tools in the study of terrain, and as such they merit close consideration here.

TABLE 15.1. Types of remote sensing techniques in relation to spectral wave length and type of ray

WAVELENGTH	TYPE OF RAYS	MAIN REMOTE SENSING TECHNIQUE
$0.3-0.7 \mu m$	Visible light	Panchromatic and true-colour photography
$0.7-0.9 \mu m$	Near infrared window (reflective infrared)	False colour and IR panchromatic
2–5 and 8–14 μm	Middle infrared windows (thermal infrared)	IR linescan imagery
5 mm–20 m	Radar bands: Q, Ka, K, Ku, X, C, S, L, and UHF	Sideways-looking airborne radar

Aerial photographs can be oblique or vertical; examples of both types are given in Figs 15.1 and 15.2. Obliques have occasionally been used for survey purposes, but their scale distortion from foreground to background is so serious that they are ill-adapted for this purpose. Their main value is as illustrations because of the graphic visual image they give of the ground. They are classified into low-level obliques taken from a viewpoint near to the horizontal, and high level obliques taken from a viewpoint near to the vertical. The former show wider sweeps of country and give a better impression of relief, but have more 'dead ground' masked by upstanding objects than do the latter.

Vertical photographs are of wider usefulness. They are taken in runs looking vertically downwards from an aircraft, which normally ensures that consecutive prints in the same run have a 60 per cent overlap ('fore and aft lap') and adjacent runs a 20 per cent overlap ('sidelap'). This means that every spot on the ground appears on at least two, usually three, and often as many as six prints, giving complete stereoscopic coverage. Because of *tilt* (the angular divergence of the aircraft from a horizontal flight path), no photograph is ever exactly vertical, but normal survey photography requires that this be under $2°$, and it is in fact almost invariably less than $1°$. This source of error as well as the scale distortion away from the centre of the photograph must, however, be borne in mind in making measurements.

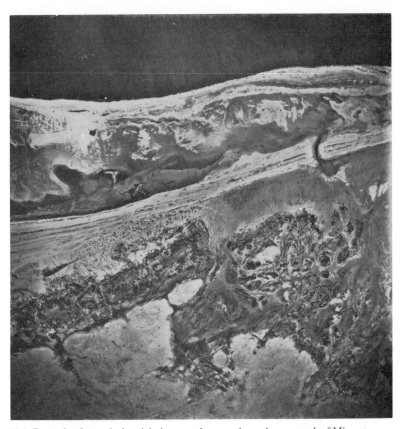

15.1 Example of a vertical aerial photograph: coastal marsh area south of Misurata, Libya. Note that this photograph reveals enough ground detail to be almost usable as a geomorphological map as it stands (Ministry of Defence, Air Force Department, photograph V:533/RAF/2684 No. 0015 of 6 February, 1964 Original scale 1:80,000 approx. Crown Copyright Reserved)

Distance measurements can be derived directly from the photograph by multiplying by the *scale factor*. This can be defined as the flying height divided by the focal length of the camera. A consideration of Fig. 15.3 makes this clear. CAO and aOd are similar triangles, therefore the distance CA on the photographs bears the same relation to the ground distance ad as does f to H.

The quality of aerial photography depends upon three factors: contrast, resolution, and parallax. The first two are self-explanatory and depend essentially on the type and grain of the film, the lighting conditions, and the quality and focal length of the camera lens. Much more information can be derived from photographs if they are viewed in pairs in stereo fashion than singly. This is because of the quality of

129

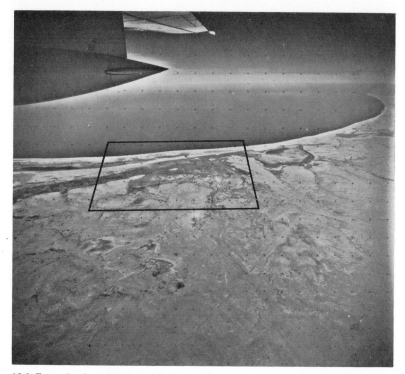

15.2 Example of an oblique aerial photograph of the same general area as can be seen in Fig. 15.1, whose approximate boundary is shown by the trapezoidal line. Note the wider spread of country visible and the more graphic impression. It would be hard, however, to use this photograph as a map (Ministry of Defence, Air Force Department photograph TRI S:543/RAF/2129 No. 0106 of 15 March, 1963. Original scale 1:60,000 approx. Crown Copyright Reserved)

parallax. It can be most simply defined as the relative movement of objects when viewed from different viewpoints and is the basis of depth perception. Everyone is familiar with the necessity of using both eyes to perceive the relative distance of objects while with one eye they appear to lie on a common plane. In just the same way, when consecutive views of the same piece of ground are taken from a moving aircraft, an observer who puts an eye at the position at which each was taken, obtains a sensation of depth. Fig. 15.3 helps make this clear.

The tool used for this purpose is the *stereoscope*, shown in Fig. 15.4. It enables the area covered on both photographs to be seen three-dimensionally by an observer viewing each through a separate eyepiece. Subject to certain errors consequent upon the imperfect verticality of all photographs and image distortion away from their centres, it enables heights to be determined and form lines to be drawn.

The determination of heights is based on the fact that, as in Fig. 15.3, the height of c above datum is equal to the flying height of the aircraft above datum (H) minus the airbase (B) times the camera focal length (f) divided by the absolute parallax of c (pc). Expressed mathematically in the terms given in Fig. 15.3:

$$h = H - \frac{fB}{pc}$$

H, h and B must be in the same units (usually metres or feet), and f and pc must be in the same units (generally millimetres).

H and f are generally shown printed on the margin of aerial photographs. B is determined by multiplying the separation of the principal points on neighbouring photographs (absolute parallax of a, or pa) by the scale factor. pc is derived by adding to pa the difference of parallax between a and c ($\triangle$ pac), i.e. $\triangle$ pac + pa = pc. The measurement of $\triangle$ pac is made with a device known as a parallax bar or more sophisticated tools based on the same principle. The essential part of the former is a pair of glass plates, each having a black mark, whose distance apart can be adjusted by means of a micrometer screw. When viewed through the stereoscope the mark can be seen to fuse to form an apparently floating object which can be moved up and down by means of the screw.

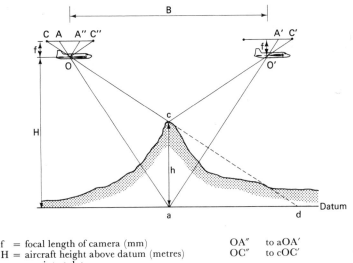

f = focal length of camera (mm)
H = aircraft height above datum (metres)
a = a point at datum
B air base (metres)

OA″   to aOA′
OC″   to cOC′

pa = the absolute parallax of a = AA″ = the air base B as measured on the photograph in mm

pc = the absolute parallax of c = CC″
$\triangle$pac = CC″–AA″ = the difference of parallax between c and a

15.3 The geometrical relations between aerial photographs and the ground surface

The height of objects can be determined by comparing micrometer readings when the floating mark is level with their tops and bases. To determine pc on Fig. 15.3, therefore, one measures the separation of the principal points of the photographs in millimetres. One then adds to this the micrometer reading on the parallax bar, determined from the difference in height between c and the principal points, assumed to represent datum at a.[1]

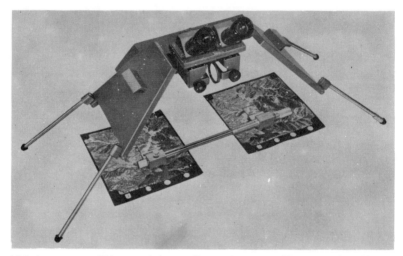

15.4 A stereoscope. This example is an ordinary mirror type, of Topcon make, with a choice of × 1, × 1·8, and × 3 magnifications. A parallax bar is resting on the aerial photographs (Photo: H. Walkland)

Scales

The scale of aerial photography used in terrain evaluation varies widely, but can be roughly divided into four ranges. Large-scale photography includes anything larger than about 1:20 000, and is useful for detailed interpretation of small features such as beach ridges, river terraces, periglacial deposits, or vegetation patterns. Because this scale requires a large number of prints to cover a relatively small area it is expensive and so tends to be used mainly in settled or economically valuable regions. Medium scale photography ranges from about 1:20 000 to 1:50 000 and is the most commonly used in terrain study. It is sufficiently detailed to reveal land elements, has a parallax distortion which is neither too small nor too great, and yet can show units as large as land systems without an excessive number of photographs. Small-scale photography ranges from about 1:50 000 to 1:80 000. Its

1. This assumption is not strictly correct, but is a useful approximation.

advantage lies in the economy of being able to cover large areas of country with relatively few photographs. It is useful for cheap and rapid reconnaissance of low value terrain such as deserts. The disadvantages are the poor resolution of ground detail and the restricted parallax. The smallest scales of all are represented by satellite photographs. These generally range from about $1:\frac{1}{2}M$ to smaller than $1:2M$ and so are separated from normal aerial photography by a wide gap in the order of scale magnitude. They are useful mainly in giving a graphic bird's eye view of the larger land units. Even areas as large as land systems are seldom readily identifiable.

Mosaics

To cover larger regions, aerial photographs are combined into *photomosaics*. These are of two main types: uncontrolled and controlled. *Uncontrolled mosaic* or *print laydowns* are composed of photographs which have not been accurately fitted into a surveyed grid. They are relatively cheap and quick to produce and are useful as rough location charts, and as a plot to the component photographs whose numbers they show. *Controlled mosaics* are based on a number of geodetically surveyed points and can be treated as almost equivalent to maps in accuracy.

Film type

There are four main types of film used in normal aerial photography: black and white (panchromatic), infrared monochrome (IR), true colour, and false colour, and each has a definite range of uses.

Black and white panchromatic film still accounts for over 90 per cent of all film used, due to its relative cheapness and availability. It is employed as the basis for topographic survey and for all normal interpretation purposes. The other types of film have specialised uses depending on their sensitivity to certain terrain features whose 'signatures' are most conspicuous in other parts of the electromagnetic spectrum.

Infrared monochrome film makes use of two facts: (1) that near-IR radiation is strongly absorbed by water, and (2) that the gross structure and condition of the mesophyll tissue of plants to it is most clearly revealed at the red and near-IR end of the visible spectrum. It is therefore better than panchromatic film in discriminating between land and water. It is especially useful for charting shorelines, the depth of shallow underwater features and the presence of water on the land surface, as for instance in channels at shallow depth underground or under vegetation. It is also valuable in distinguishing between different types of vegetation and identifying areas of plant disease where no photosynthesis is taking place. IR film also has the advantage of a greater capacity to penetrate haze than does conventional photography, because it senses at the slightly longer wavelengths which are less subject to scattering by atmospheric haze particles.

The use of true colour photography is rapidly increasing today because it has certain marked advantages over panchromatic. Because it displays variations of hue, value, and chroma rather than tone only, it normally provides much more refined imagery. Its inherent superiority was demonstrated by Cooke and Harris (1970) in a study of the Isle of Man which showed that small individual features could be identified more readily. It does, however, suffer from certain technical limitations when compared with panchromatic film. Apart from greater cost, these include the narrower range of exposure that will provide an acceptable image, the more critical film processing requirements, and the greater difficulty in duplicating colours in different parts of the same picture and on different prints from the same film roll.

False colour is perhaps the most versatile and potentially valuable type of film for terrain interpretation purposes. Apart from the uses and advantages listed above under both IR monochrome and true colour film, it also provides a more sensitive means of identifying exposures of bare grey rocks than any other type of film because such exposures appear in shades of blue which depend to some extent on the freshness of the exposure. But it does suffer from the technical limitations noted above under infrared and colour photography.

A recent development has been the adoption of systems of *multiband sensing*. These attempt to record simultaneously signals from several bands of the spectrum, using a *multiband* camera. One such camera records, simultaneously, information from nine portions of the visible and near-IR parts of the spectrum. It consists of nine matched lenses with different filters, some using panchromatic and some black and white IR film. It can be supplemented by extra cameras holding true and false colour film, and other remote sensors for longer wave bands. It provides much more information than could be obtained from any single photograph.

Passive indirect systems

Passive indirect systems sense IR radiation at longer 'thermal' wavelengths beyond the range of the visible and near-IR spectrum. They do so through 'windows' in the IR spectrum that allow terrestrial radiation to escape through the atmosphere with least loss of absorption. The most important of these windows are those between approximately 2 and 5 μm and between 8 and 14 μm.

The sensitivity of detectors, and the wavelengths to which they respond, vary widely, but the most important for terrain interpretation is the technique known as *infrared linescan* (IRLS). Its basis is the translation of ground radiation into electrical signals which can be transformed into visible light via a cathode ray tube and are then recorded on film or magnetic tape. The technique involves scanning a succession of parallel lines across the track of an aircraft with a scanning spot a few metres across. The spot goes back and forth in such a way that no gaps are left between its consecutive passes. As it records only an average radiation, the limits of resolution depend on the size of the spot. Its diameter is normally of the order of 2 to 3 milliradians, i.e. if the plane is flying at 1000 m, the spot will be 2 to 3 m across. The radiation received in the aircraft is reflected by a rotating parabolic mirror on to a detector which generates signals which vary in intensity with the amounts of incoming radiation. These signals modulate an electron beam whose projection on to a cathode ray tube enables a continuous black and white photographic print of the ground to be obtained. This print tends to be increasingly distorted with distance from the line of flight, limiting the total useful angle of scan to about 60° on each side of it. A diagram of a simple line scanner is shown on Fig. 15.5.

Multispectral line scanners are a refinement of this method, differing from it in one important respect. Instead of having a single detector mounted at the focus of a parabolic mirror, they employ a dispersing system which produces a focused spectrum of energy. By correct placement of an array of detectors, the dispersed signal can be recorded in

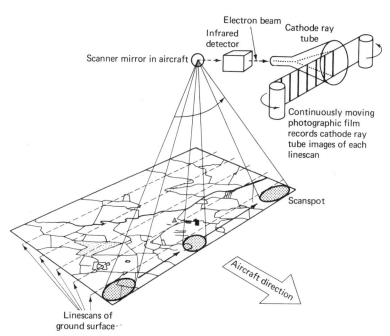

15.5 The principles of obtaining infre-red line-scan imagery (Source: R. V. Cooke and D. R. Harris, 1970, p. 8)

narrow, discrete spectral bands, or channels. Twelve, eighteen, and twenty-four channel scanners are in use. This is as yet almost untried in terrain evaluation but its potential may be considerable as it allows non-relevant data to be eliminated and discarded before producing an image, something which is not possible with camera systems (Olson, 1970). A diagram of a multispectral line-scanner attached to a data processor is shown as Fig. 15.6.

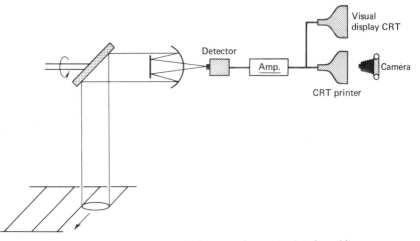

15.6 The system for the management of information from a simple infra-red line-scanner (Source: C. E. Olson, 1971, p. 12)

Active indirect systems

Whereas passive sensors depend on solar energy and record reflected and emitted radiation, active sensors, which operate at longer wave bands, generate their own energy. The most important of these for remote sensing of the earth is sideways-looking airborne radar (SLAR). Short pulses of energy in a selected part of the radar waveband are directed sideways to the ground from transmitting antennae on both sides of an aircraft. They strike the ground along successive range lines and are reflected back again at time intervals related to the distance of the ground from the aircraft. The returning signals are transformed into black and white photographs via a cathode-ray tube using a technique similar to that used in IRLS. One limitation is that return pulses cannot be accepted from any point within 45° from the vertical, so that there is a blank space under the aircraft along its line of flight, and there is also increasing distortion of the scale of the image towards the track of the aircraft, the reverse of that which effects IRLS. A diagram of the method is shown as Fig. 15.7.

As with camera and line-scanning systems, recent developments in radar have been toward multiband systems which use a number of different wavelengths in combination rather than a single one. Although such systems have not been used extensively, the differences in soil and vegetation penetration capabilities in different bands suggest that some useful applications may be forthcoming. Further development also seems likely of *multiple polarisation* and *coherent* radar systems. The former have the extra capacity of distinguishing objects by their different degrees of polarisation, while the latter improve resolution of images of objects to the point at which they are essentially independent of their distance away.

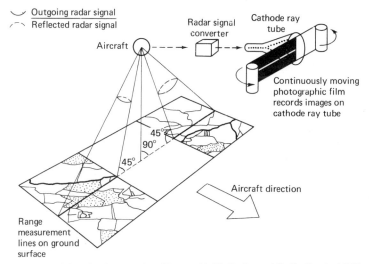

15.7 The principles of radar scanning (Source: R. V. Coche and D. R. Harris, 1970)

Another development has been the use of *lasers*. The Aero Service Division of Litton Industries (Anon, 1965) have developed a device by which an airborne profile recorder, used in conjunction with a laser, can give an accuracy of less than 30 cm in recording micro-relief from an aircraft flying at 400 km.p.h. at an altitude of 305 metres.

Finally, methods have now been evolved for 'sniffing' the ground surface to detect the presence especially of certain types of organic waste.

Evaluation of remote sensing imagery

Progress has been made in the past fifteen years in automated picture interpretation of aerial photographs. This involves three steps:
1. *pre-processing* which transforms the original picture into a new one enabling quantitative information to be derived from individual points or 'unit squares';

2. *quantification* of these points;

3. *evaluation* of the pattern of quantified points in relation to predetermined classes (Rosenfeld, 1968).

It is now also possible to scan imagery directly and automatically by means of photoelectric scanners, to encode the information and to record it in terms of geographically located *x-y* coordinates. This can even be done directly from the sensing device so that the need for photographic imagery is obviated altogether. Multispectral line-scanners eliminate the problem of getting congruent image registration inherent in camera systems, because all the data channels feed into a single optical system rather than a number of independent lenses. They also have the advantage of having less image distortion than radar.

Nevertheless, the end products of all forms of remote sensing generally appear as visible photographs and so essentially the same processes of interpretation are employed in all. There are, however, certain features about the 'photographs' derived from IRLS and SLAR systems which give them distinct values and uses in terrain interpretation.

IRLS sensing depends essentially on clear and calm weather conditions. High winds and humidities or a polluted atmosphere can lead to blemished images. Although capable of recording differences of as little as $0.1°C$, these do not show differences in absolute ground temperature, but only in radiation emission and so require careful calibration before absolute values can be obtained. Furthermore, thermal emissions vary so much through the day that time must be carefully considered. IRLS is most useful for the study of direct thermal phenomena such as forest fires, ice, snow, hot springs, and situations where waters of contrasting temperatures meet. While valuable for certain purposes, it seems likely that the most important use of IRLS will be to supplement more conventional techniques by providing information on smaller features which are characterised by distinct temperatures.

Despite the technical limitations already mentioned, SLAR has certain virtues. Most notable in its long-range capability, its independence of lighting and atmospheric conditions, and the possibility of discovering the height of objects. Although there is some evidence that the use of sophisticated equipment may permit discrimination among different types of vegetation, it appears that the method is mainly useful for broad reconnaissance or emergency observations of terrestrial phenomena at night or when they are cloud-covered.

As Olson (1971) has pointed out, the use of multispectral scanning exacerbates what is fast becoming one of the chief problems of the environmental research process: the superabundance of data, most of which is not relevant to the questions to which an answer is sought. A human interpreter can integrate and interpret only about three separate images simultaneously. When given more than three, he selects, consciously or unconsciously, the three which he feels give him the most information and concentrates on these. His capacity cannot be effectively increased beyond this, so that to interpret more than three, he requires some sort of automated decision-making machine to reduce the amount of information he must process. The best way of achieving this is to make sensors more selective so that they can discard non-relevant information before producing imagery. In the long term, this will tend to rule out the use of normal camera techniques and replace them with a uniform method of scanning the multispectral band used.

The most up-to-date approach to terrain evaluation involves the use of rockets and satellites, although the development of these as refined research tools still lies in the future. Both rockets and satellites view the ground from greater altitudes than are possible for aircraft, but suffer from the consequent difficulty of obtaining adequate resolution of ground detail. Rockets have the advantage of being considerably cheaper in that they do not require to be part of a costly space programme. For this reason, they are more suited to the needs of developing countries.

An example of a rocket equipped with a camera system for ground viewing is the UK *Skylark*.[1] Its first launching for terrain study was at Woomera in March 1972, and further firings are scheduled in Argentina in early 1973 where they will be used in conjunction with balloons at 40 000 metres and aircraft at lower levels, both carrying cameras. This rocket has a trajectory about 200 km long and reaches a height of about 300 km at apogee. It is programmed so that horizon sensors stabilise it and rotate it in a horizontal position and allow it to take a series of high oblique photographs with two or more cameras which give a panoramic view of the ground about 800 km in diameter. The photographs are directly on film and so are of better quality than can be obtained from television systems, but it is not yet possible to make a final determination of the value of rockets as terrain-sensing platforms.

The potential of satellites in remote sensing has led the United States National Aeronautics and Space Administration to include the launching of an earth resources technology satellite in their space programme. The first of these (ERTS A) was in mid 1972 with a second (ERTS B) to follow about a year later. ERTS A will circle the earth in a sun-synchronous polar orbit which will enable it to pass over every spot about once in eighteen days. It will carry three television cameras, two sensing within the visible spectrum and the third within the near-infrared wave band and a 4 channel multi-spectral scanner. ERTS B will add a capability in the middle infrared range. Both satellites will yield a continuous stream of data on magnetic tape and as photographic images. This should give a resolution down to about 50 metres, but even if this is not attained in practice, the satellites will give greater breadth and frequency of coverage than is possible with aircraft, and will be sensing at a scale which is still useful for terrain as well as geological, topographic, and hydrological surveys.

1. Designed by the Royal Aircraft Establishment, Farnborough.

16
Landscape sampling

Terrain is continuous, and if we are to discover its properties, at least those of its materials such as soil and rock that lie below the surface, then our observations must be restricted to a small part only, that is, to a sample. It will also be clear by now that terrain is very variable. Although by classifying the terrain we aim to reduce the variation with which we have to deal at any one time to manageable proportions, the variation remaining in each class is nevertheless still much greater than is introduced by observational error. Samples differ therefore from one another, and worthwhile information can be obtained only when several, and possibly many, sampling sites have been observed. Replication alone, however, is not sufficient to ensure reliable information. We must also avoid bias, because information from biased sampling is often misleading. To avoid bias samples should be chosen so that initially every site has an equal chance of inclusion.

The sampling of terrain differs from that given for vegetation in chapter 7 in being based even more completely on locational and spatial criteria. This is because terrain is more unalterably fixed and less ephemeral than is vegetation, and is without the limited degree of independence of environment possessed by all living things.

Density and depth of observation sites

Certain empirical standards have been developed for the density of site observations by such organizations as the United States Department of the Interior's Bureau of Reclamation (1951) and land development consultants such as Hunting Technical Services Limited (1958). These can be taken as roughly, though not necessarily exactly, applicable in most parts of the world, especially developing countries. Reconnaissances require a minimum of 1 site per 1 250 hectares, although 1 site per 250 hectares should be regarded as normal. Some detailed surveys concerned with the feasibility of irrigation and other development projects require about 1 site per 100 hectares, and detailed surveys require between 1 site per 25 hectares and 1 site per 10 hectares depending on the complexity of the landscape and the size of the financial investment involved in its development.

Soil observations are made to a depth adequate to penetrate any horizons present. This depth varies, but for areas considered for agricultural development, especially where irrigation is envisaged, 1·5 metres is normal. This approximates to the maximum rooting depth of most annual crops and to the depth to water table above which, in most soils, there is potentially a continuous capillary film reaching to the ground surface. A certain proportion of observations, however, are made to 5 metres in order to determine the likelihood of development problems arising from the occurrence of underground rock or bands of gravel, clay or gypsum. At least one such deep boring is done per 1250–2500 hectares in all types of survey, but especially in areas where subsoil conditions cannot be inferred from the geology e.g. alluvial plains.

Distribution

There are two basic approaches to the problem of subdividing an area in order to distribute traverse lines or sample sites over it. Where the landscape is complex or little known, all preliminary judgements about it are likely to be questionable, and it is necessary to select traverse or sample sites at random over the whole area. This approach has the additional recommendation that the mathematical techniques which must be used to analyse the field data work most effectively and give conclusions of the widest validity if the sample is random.

Alternatively, one can seek to exploit existing knowledge about the landscape to the utmost and to subdivide it on climatic, geomorphic, and general scientific grounds as far as possible, only falling back on randomization when these resources are exhausted. This reduces the range of mathematical techniques which can be employed and the scope of the conclusions which can be derived from them, but gives more scope for the use of common sense and experience in field observation.

The choice invariably lies somewhere between these two extremes and must be determined from a number of considerations, the chief of which are the scale of the survey, the complexity of the landscape, the amount of information already available about the area in question, and the degree of detail required about it. On the whole, where the survey area is large or relatively well known and covered by detailed maps and aerial photographs, major physiographic distinctions will be more obvious and better understood, and sampling will be stratified in accordance with them. Where these conditions are not fulfilled, simple random sampling will be favoured.

In the normal situation, therefore, an area is stratified into its obvious physiographic regions, subdividing each into facets as small as possible with the tools and information available. A decision will previously have been reached, based on considerations of time and cost, of the overall amount of sampling to be done. The sample sites are then distributed among the facets in such a way as to represent each one adequately but to concentrate effort on those with the highest economic potential or natural complexity.

It is sometimes necessary, for research purposes, to obtain a sample of large areas of the earth's surface for study. This is best done by following these same principles: first, a stratification of the area involved into regions on the basis of existing knowledge; second, a classification of these regions into physiographic types; and third, a selection of samples of each of these types and of sample areas within these by a method of randomization which may follow practical convenience so long as this does not involve biasing the choice of areas by factors which are inherent in the terrain itself. An example of one such scheme is given in Appendix A.

The actual distribution of sites within each of the facets into which the area has been stratified may be random, grid, or selective. The random method selects points at random intervals along randomly distributed traverse lines, or alternatively divides the area into blocks and then chooses which to sample from a table of random numbers. The advantage of this method is that statistical techniques of interpretation are most widely applicable to such data. The grid method distributes sites evenly over an area and thus makes their choice less completely random. Its advantage lies in bringing every part of the area into the closest possible proximity to at least one observed site. It resembles random sampling, both in being easily amenable to statistical treatment and in taking no advantage of the easily ascertained divisions of the landscape. Moreover the locations of grid borings may overrepresent simple or unimportant parts of the terrain which happen to be areally predominant. The selective method continues the original stratification by attempting to distribute sites according to inferred geomorphological criteria within the sampling units. In practice, however, solutions of such mathematical neatness are seldom attainable under field conditions and it is generally acceptable compromise to choose sample sites within the facets on the basis of their general accessibility, provided that the choice of their location is biassed as little as possible by the nature of the terrain itself. In this way they can be considered adequately random.

Each unit or facet is then sampled with enough sites to give an indication of the variance of measured properties. In general, this means at least three or four sites per occurrence.

It is sometimes necessary to verify whether a particular sampling policy has covered the full range of possible terrain variety within an area and that no 'gaps' have been left. This may be achieved by a method such as the one illustrated on Fig. 16.1. In essence it involves the construction of a three-dimensional diagram whose axes represented the main types of natural variety being sought: in this case the lithology, texture of surface materials, and geomorphic form of world deserts. The

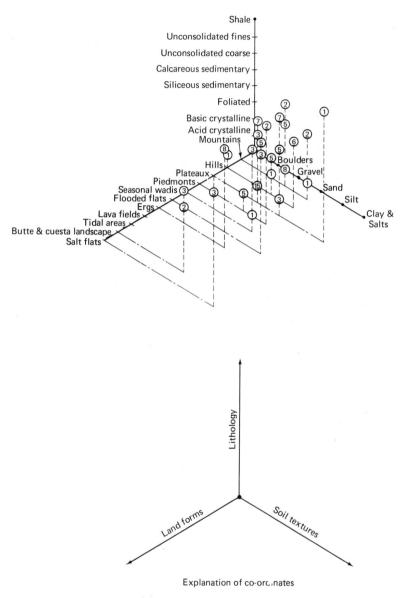

16.1 Three dimensional graph used to assess how far all combinations of the three variables: surface material, lithology, and geomorphic form, had been represented by a number of observed examples (Source: R. M. S. Perrin and C. W. Mitchell, 1970, p. 234)

143

known examples are plotted on to the diagram. This then reveals the presence of clusters, where data is abundant, and gaps where more data needs to be sought. Fig. 16.1 shows, for instance, the abundance of examples of mountains and plateaux which occur on acid crystalline and consolidated sedimentary rocks, the dominantly coarse texture of surface materials and the absence of examples of lava fields and tidal areas.

17
Field techniques

The field work required in integrated survey differs from that required for specialist surveys in the wider range of data collected. It involves the whole range of basic geological, geomorphological, pedological, ecological, and both actual and potential land use observations, either carried out by a team of specialists travelling in company, as is the CSIRO practice, or by a single 'integrated surveyor'.

The field observations are based on a number of site visits at which pits or borings are made and samples taken. Chapter 16 showed that the density and distribution of the sites and the depth of the borings depend on the amount of prior knowledge available, the degree of detail of the survey, and the expected complexity of the landscape. General decisions have been reached about these before the field work is started.

A work plan is made for the field observations. Once the study area has been defined, it is carefully scanned on the maps and aerial photographs to find routes and plan the daily field traverses. These aim to cover the area in blocks which do not leave isolated gaps requiring special visits to complete.

The activities to be carried out at each site are limited to those which are relatively quick and accurate under field conditions and which do not require equipment which is either complicated, heavy, or cumbersome to transport, easily damaged by shocks or dirt, or needing too high a standard of skill, sharp eyesight, manual dexterity, or patience to operate.

Site observations can be considered as falling into a number of distinct groups: (1) landscape and mesorelief, (2) ground surface and microrelief, (3) soil profile, (4) field tests, (5) soil and water sampling, (6) vegetation observation and sampling, and (7) land use observation. Each of these aspects has its own techniques and most involve the use of special tools. Each site in theory represents all the land up to the midpoint to the next site in any direction. In practice, however, observations are based on an 'envelope' with a relatively arbitrarily chosen radius, usually about 100 metres, centred on the site. Neighbouring sites should not be nearer to one another than the diameter of this envelope.

It is useful to have comprehensive proformas for recording information. These may be in the form of cards to be punched at the study sites, provided they are not too difficult to use or vulnerable to damage under field conditions. But it is more common to write field notes on printed sheets of the type illustrated on Fig. 17.1 and then abstract the data on to a different format for storage purposes.

Landscape and mesorelief observations can be derived to a large extent from maps and aerial photographs before visiting the site. They have three main aspects: relief, landscape aesthetics, and erosion risk.

Observations of relief must attempt an interpretation of the basic geology and landscape evolution and specifically include direction (aspect) and angle of maximum gradient, maximum relief amplitude within 100 m of the site, and the proportion of the total area occupied by bare rock or gullies. Landscape assessment is best done by a combination of the *isovist* method of defining landscape units with a scaled estimate of landscape quality within a 360° circuit of the site with high values for vertical contrast, pleasing colour patterns, and the presence of water.

The assessment of erosion risk properly includes all aspects of land conservation. Observations are made of slope and gully stability and the risk of deposition, of windborne or waterborne debris or poisons, either from sudden storms or from the slow pollution of air or water.

Ground surface observations include microrelief and the character of surface materials. Microrelief is often hard to assess quantitatively and must generally be based on a verbal description of the ground surface unless either a morphometric map or a detailed study of a sample area is being undertaken. Stone and Dugundji (1968) describe a measuring device, called the 'continuous ride geometry vehicle' which can be used for recording a microrelief profile. As the vehicle traverses the terrain, it gathers, edits, amplifies, and digitises data on x and y coordinates yielding a magnetic tape suitable for use in a digital computer. Such tools are, however, expensive, and not always justified by the information they yield.

The character of the materials which constitute the land surface is also recorded. These constitute that part of the regolith which occurs at depths accessible for quarrying or open-cast extraction, and include ore bodies, building stones, constructional gravel or aggregates, lime, china or brick clays, etc. The character of the ground surface is also recorded in detail. Where not covered by vegetation, it is divided into proportions occupied by stones, sand, and silt or clay. Assessment is then made of the lithology and the maximum and average size of stones, the proportion of sand in the form of drift and hummocks, and the height of the latter; and, for the finer material: the dimensions of the crust, if present, and the size and spacing of cracking. If possible, a full vegetation survey should be carried out along the lines outlined in chapter 7, but it may be necessary to reduce this to such essentials as merely recording percentage ground cover and the heights and spacings of dominant plants.

A detailed description of the methods of soil profile recording are given in such books as the USDA *Soil Survey Manual* (Soil Survey Staff, 1951) and Clarke's *The Study of Soil in the Field* (1957). The basis is the systematic recording of soil profiles by horizons. Each horizon is assessed for texture by the feel of the moistened soil, and then by visual observation: the percentage abundance, lithology, maximum and average size of stones, Munsell colour, moisture, consistency, structure, cracks, roots, and visible salts.

As many soil samples as possible are taken for laboratory analysis to support the survey, but it is usually necessary for reasons of economy to confine the sampling to a limited proportion of the sites and horizons only. Samples are either of disturbed or undisturbed soil. The former are obtained simply by using an adze and scoop on a profile face or by emptying material brought up by the auger from each horizon into bags. Undisturbed samples are obtained by driving metal tubes with sharpened ends into the ground surface or steps at specified depths in soil pits and then sealing the ends with rubber caps before placing in polythene bags. Water samples are taken of any shallow ground water in pits, borings, or well sites. The laboratory analysis of these samples is considered in the next chapter.

There are two other types of field test which are important for terrain surveys. One of these is directed towards the assessment of the capacity of the ground to sustain the passage of vehicles or to serve as a foundation for buildings or roads. Vehicle passage is dependent on soil strength, shearing resistance, and 'bumposity'. Tools have been designed to measure all these characteristics, though only the first is widely enough used and understood to be worth while assessing routinely.

A number of penetrometers and shear vanes are available for measuring soil strength in the field. Penetrometers can be pocket-sized or haversack-sized. The most satisfactory seem to be the USAEWES (1963c) or MEXE (1968) patterns. These are similar except that the latter is housed in a stronger casing and is more robust. They work on the principle of recording on a dial the resistance of the ground to the insertion of a rod of known cross-sectional area, operated manually by pressing on a handle. Readings are taken as the rod passes marked depths on the rod. The dial is calibrated in lb/square inch and California Bearing Ratio (CBR) units, which MEXE (1968) correlate with vehicle performance. The number of readings required at each site and the statistical methods for handling the results have been worked out in some detail by Beckett and Webster (1965d).

Shear vanes measure the resistance of the soil to a twisting motion of a vertical plate, but have not proved able to give sufficiently reproducible results to be widely used.

Empirical tests of terrain trafficability have been worked out using an instrumented land rover (MEXE, 1966). These aim to compare the

(a) TERRAIN SURVEY PROFORMA: SOIL PROFILE

Depth	Colour	Mottling	Texture	Material	% of soil	Av. size	Max. size	Shape	Salts %	Moisture and Consistency	Structure, Cracks, Pores, etc.	Roots

REMARKS

Type of Sample	Depth	Sample No.	E.C.

Location

Sortie — Print

Surveyor

Water Table

Bore
Pit

Field — Final

Facet class

Mosaic

Date

Time

148

(b) TERRAIN SURVEY PROFORMA: GROUND SURFACE

Date & Time	Surveyor	Sortie	Print	Facet	Site No.

GEOLOGY

Aspect	Slope form (Convex, Concave, Linear)
Gradient up:	Slope changes
Gradient down:	

MICRORELIEF

Bare rock	Channels
Undulations	Gullies
	Hummocks

Diagram & Remarks

VEGETATION

% Cover   :

Height of dominants

Spacing of dominants   :

Species:	% of total
Dominant	

Secondary

SURFACE

Colour

Stones

Material

Size

Shape

Spacing

Cone Positions	0	3"	6"	9"	12"	15"	18"	21"	24"

Sand

Sheet/drift

Hummocks

Ripples

Silt & Forms

Clay & Forms

Type of Cone

17.1 Examples of field proformas for use in terrain surveys
  (*a*) for soil profile observations
  (*b*) for ground surface observations
  (Source: R. M. S. Perrin and C. W. Mitchell, 1970, pp. 286, 288)

overall effect of the terrain in resisting the rolling resistance of the vehicle with the effect of a theoretically ideal road surface. They therefore give a resultant of the combined effects of slippage (lost traction) and sinkage (poor weight support) without distinguishing between the two. The method is to express the distance travelled across selected tracts of terrain as a percentage of that travelled in the same time along a level stretch of road, calling the result the *going coefficient*.

The vehicle used in the original research was a land rover, which was held in constant gear and maintained at constant engine speed with the aid of an impulse tachometer to show the number of engine revolutions per minute and a throttle screw to control them. The distance travelled was measured by a revolution counter on a trailing bicycle wheel which was fitted with a spring to hold it down on the ground. In studies in Libya, Bahrein, and the Trucial States, it was found that most terrain types gave values of the going coefficient between 70 per cent and 90 per cent and that immobilisations usually occurred when these fell below 60 to 70 per cent, (Perrin and Mitchell, 1970). As might be expected, wet saline ground and sand dunes scored lowest, and dry, firm, desert plains highest. A complete empirical assessment of trafficability should also include a measure of bumposity, i.e. the frequency and size of bumps encountered, and the distance of diversion necessary round obstacles. Experimental tools such as bump counters have been designed to measure the first (MEXE, 1966), and the second has been assessed by measuring the percentage of additional distance imposed on a vehicle by the microrelief when driven between two intervisible points of predetermined separation in Yuma, Arizona (USAEWES, 1962), and in the Trucial States (Perrin and Mitchell, 1970).

The intrinsic capacity of the soil material to transmit fluids is important in any study involving hydrology or the movement of water in soils. This parameter is known as the *hydraulic conductivity* and must be measured in saturated media where flow is laminar and the gravity factor is eliminated. The best method is the pump-out auger hole method described by van Beers (1958).

The equipment for this test includes a post-hole auger with sufficient extension rods to penetrate well below water table depth, a baler, a stopwatch, and an electrical probe fitted to a galvanometer. Where the soil is likely to collapse, it is necessary to carry perforated metal tubing to case the holes. The baler is a section of metal tubing with a bottom opening on a hinge to admit the water and the other end threaded to take auger rods. The electrical probe consists of an electric plug on a wire connected to a galvanometer powered by a battery, which gives a flick of the galvanometer needle whenever the plug makes contact with the water surface as it rises.[1] The field procedure is rapidly to bale out the auger hole and measure the exact time required for the water table

1. Designed by Hunting Technical Services Limited.

to rise in the hole by raising the probe regular amounts measured along a scale of marks at 1 cm intervals. The items of equipment used in the test are illustrated in Fig. 17.2. The result are calculated according to the formula given in Fig. 17.3.

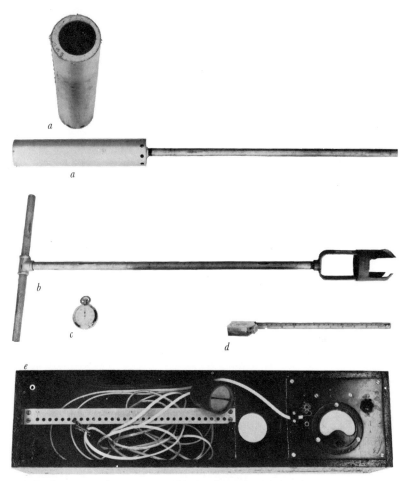

17.2 Equipment used in measuring soil hydraulic conductivity by the pump out auger hole method:
  (*a*) baler
  (*b*) 4 inch Jarrett post-hole auger
  (*c*) stopwatch
  (*d*) tape measure
  (*e*) 'Δy box'. The plug drops through the white hole down into an auger hole and records the depth of water table by a flick on the galvanometer. The small clip fits into the small holes allowing the wire to be pulled up in exact units of 1 cm
  (Photograph: H. Walkland)

Where the water table is below augering depth, i.e. below about 5 metres, it is not possible to make a field measurement of hydraulic conductivity of the soil, and it is necessary to fall back on empirical measurements of its *surface permeability* or *infiltration rate* which can be calibrated against hydraulic conductivity measurements where these are available. Two types of pour-in tests can be employed to determine infiltration rates: the *pour-in auger hole method* and the *double cylinder method*.

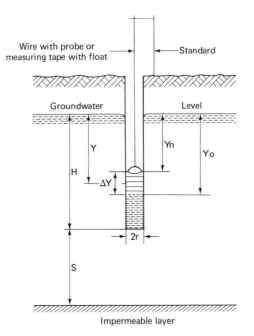

Wire with probe or measuring tape with float — Standard

Groundwater   Level

Y Y_n Y_o

H ΔY

$2r$

S

Impermeable layer

H = depth of the hole below the groundwater table.

y_o = distance between the groundwater level and the elevation of the water surface in the hole after removal of water at the time of the first reading.

y_n = the same at the end of the measurement. Usually about 5 readings are taken.

$\Delta y = \Sigma \Delta y_t = y_n - y_o$, the rise of water level in the hole during the time of measurement,

y = distance between the groundwater level and the *average* level of the water in the hole during the time of measurement.

$$y = \frac{y_n - y_o}{2} = y_o - \tfrac{1}{2}\Delta y.$$

r = radius of the hole.

S = depth of the impermeable layer below the bottom of the hole.

Note: The measurements should be completed befor $y_n < 3/4\ y_o$, *or* $\Delta y > 1/4\ y_o$.

17.3 Diagram and formulae illustrating the method for measuring soil hydraulic conductivity by the pump-out auger hole method (Source: W. F. J. van Beers, 1958, p. 12)

The pour-in auger hole method is described by Hunting Technical Services (Republic of the Sudan, 1963). A dry auger hole is bored through the horizon to be measured, which should be at least 25 cm thick, making sure that there are no cracks. The hole is packed with gravel and water is run in and maintained at the level of its top. When inflow rate has stabilised, it is measured by catching the flow in a graduated cylinder over a timed period. Infiltration rate is calculated from the following formula:

$$I = \frac{864Q}{Cu\ rh}$$

Where:

I = permeability in m/day per unit hydraulic gradient.

Q = rate of steady inflow in ml/sec.

r = radius of bore in cm.

h = height of water column in cm.

Cu = an empirical coefficient of conductivity which must be assessed from h/r by interpolation between 32 when $h/r = 10$ to 48·5 when $h/r = 20$.

The double cylinder method is shown on Fig. 17.4. It uses two lengths of open-ended cylindrical steel pipe, sharpened at one end, and with diameters not less than about 15 cm and 25 cm respectively. These are driven a short way into the ground concentrically and both are filled with water. Measurement is made in the inner one only and it must be somewhat arbitrarily assumed that the water it contains penetrates vertically downwards into the underlying soil column. A constant head is maintained by arranging that the supply is from an outlet tube under an inverted bottle. Infiltration rate is calculated by dividing the quantity of water lost from the bottle in a given time by the cross-sectional area of the inner cylinder.

Fieldwork in terrain surveys cannot be considered complete without interviewing and questioning local landowners to obtain as much information as possible about land uses, crop yields, and local economic factors in general. Where possible, this is best done by using printed questionnaires, an example of which is given by Soil Survey Staff (1951).

The final phase is post-field work. This includes the preparation, listing and submission of soil and water samples for analysis and of plant samples for identification. Field data are tabulated and punched onto cards or entered on to base maps.

Laboratory analyses[1]

There are a number of relatively simple laboratory tests of the physical, chemical, and mineralogical properties of soils important to agriculture

---

1. The author is indebted to Dr R. M. S. Perrin for this section.

and engineering. These are based on analysis of soil samples taken in the field. They are described in modern laboratory textbooks, so no attempt is made here to do more than mention the most important.

The main physical parameters requiring measurement are *particle size distribution* (mechanical analysis), *Atterberg limits* (i.e. the *plastic limit*, the *liquid limit*, and the *plasticity index*), *linear shrinkage, shear strength, compressibility, dry density, porosity* and *pore size distribution*, and the *moisture characteristic* including *field capacity* and the *permanent wilting point*. Methods are given by Means and Parcher (1964), Black (1965), and British Standards Institution (1967).

The chemical properties most usually required for agricultural purposes are *soil reaction* (pH), *salinity* (usually measured in terms of electrical conductivity), and sometimes *sodium absorption ratio, total carbonate, 'available' phosphorus* and *potassium, exchangeable cations,* and *cation exchange capacity. Total nitrogen* and *organic carbon* and some trace elements such as *boron, copper,* and *molybdenum* are occasionally important. Methods are given by Black (1965).

Finally, the determination of the mineralogy of soil clay fractions (<2 micron effective diameter) gives a clue to pedogenesis and helps to account for some of the physical and chemical characteristics of soils. Methods are given by Brown (1961) and Grim (1968).

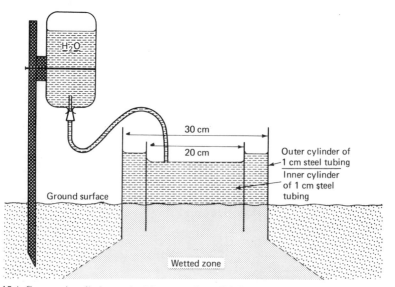

17.4 Concentric cylinder method for measuring soil infiltration rates

18
An introduction to the literature on statistical methods

Having suitably sampled the terrain and made whatever observations are appropriate (see chapter 17 for a description of these) we shall generally wish to examine the data derived so that we can make statements about the terrain they represent with reasonable confidence. This is the purpose of statistics. The subject of statistics is very large, and there are many techniques from which an investigator may choose. However, choosing the correct or best technique in any given situation can be difficult; there are many pitfalls to trap the unwary. The beginner should always consult a qualified statistician, first to ensure that his sampling scheme is likely to produce the information that is required, and second to choose methods of analysing the data when he has them. It is good practice to write down algebraically every step in any analysis that is to be performed before the survey is undertaken. Unforeseen difficulties can arise during survey, and the investigator should be prepared to modify the analysis as a result.

It is quite beyond the scope of this book to describe in detail any of the techniques that have been found profitable. To help the reader, however, those statistical measures and tests that he is most likely to find useful are mentioned, and standard statistical works referred to. Several technical terms have had to be used without definition, but they are underlined so that they can act as signposts to the statistical literature.

Statistics are required for at least three purposes in terrain evaluation. In the first place they provide numerical description. The arithmetic *mean* is familiar enough as the measure of central tendency or 'average' in a sample. However, in some circumstances it can be misleading, and better measures might be the *mode*, the most frequently occurring value, or the *median*, the middle value of those observed. Other descriptive statistics that are often required are the *variance*, which describes the spread of values in the sample from the central value, and *skew*, the extent to which this spread is unequal about the centre. When observations are of the kind 'presence' or 'absence' of some character then the *proportion* of those that possess the character is usually needed.

The numerical values of these quantities properly refer only to the sample from which they were derived. If we wish to use them to describe all the terrain from which the sample was drawn then we must realise that they are subject to *sampling error*. We need to know what *confidence* may be placed in the sample values as estimates of the terrain studied. In order to calculate the confidence an investigator often makes certain assumptions. In most simple applications he assumes that the variation within classes is random; i.e. the sampling error associated with one observation is quite independent of other observations. He often assumes that these errors are *distributed* in a certain way. If the data are 'presence' and 'absence' then they are usually assumed to have a *binomial* distribution. If they are measurements they are often assumed to have a *normal* distribution. Statistics are most powerful when applied to normally distributed data, and it is always worth examining measurement data to see whether they are normally distributed before proceeding to calculate confidence levels. Data that are not normally distributed can sometimes be transformed so that they become normal.

Statements of confidence take the form: 'On the sample evidence there is k per cent probability that the mean value of the property of interest of this terrain lies in the range x_1 to x_2', or 'The mean value of this terrain property is $\bar{x} \pm c$, where $+c$ and $-c$ are the symmetrical $(100 + k)/2$ per cent confidence limits'. One of the main purposes of terrain classification and evaluation is to predict conditions at unvisited sites. The best estimates of conditions will usually be mean values, and they should always be accompanied by the associated statements of confidence. Translated into operational terms they can take the form: 'There is only one chance in ten that the strength of the soil is less than y, and that a vehicle of a given axle load will sink.'

The third important use of statistics is for comparing two or more classes of terrain. Comparisons may be needed to decide which kind of terrain would be best to develop for agriculture or where would be best to exploit the soil for road making materials. Alternatively a surveyor might not be altogether happy with his classification and might wish to know whether it was worth trying to distinguish two classes of terrain. We assume that sampling evidence suggests that there are differences sufficiently large to be worth considering. Such comparisons involve statistical *testing*.

To make a test the statistician acknowledges that there is sampling error, and that he cannot be sure whether any differences he observes between classes of terrain are more than sampling variations. He therefore adopts the following position: If there are no real differences between the classes being considered, what are the chances that the differences observed could have arisen by chance? The situation being tested is often referred to as the *null hypothesis*. Conventionally the null hypothesis is accepted in a single test if the chance of observed differences arising is better than 5 per cent, and rejected otherwise. If the null hypothesis is rejected the differences between classes are said to be significant. If an investigator is unhappy about proceeding on this basis he can apply a more stringent test by choosing to work at the 1 per cent,

0·1 per cent or other level. Alternatively he can take the view that the evidence suggests the differences are real but that he would like to be more sure. To increase his confidence he can carry out further sampling. If there are real differences between classes then increased sampling will increase the chance of finding them; if there are no real differences increased sampling will decrease the chance of differences appearing in the sample evidence.

Testing for differences of proportions in samples is usually done using *chi square* (χ^2). Testing for differences between means of measured properties is usually done by classical *analysis of variance*, or, if there are only two classes, the even older, *Student's t*. Both of these assume normal distributions, but are robust against departures from normality, i.e. they are unlikely to mislead over a wide range. More recently several methods have been developed to compare sample means without making any assumptions about the distribution. They are known by their authors' names, and among the more important are the *Mann-Whitney U test*, the *Kolmogorov-Smirnov tests*, and the *Kruskal-Wallis* analysis. They have less power than the classical methods for normally distributed data, but can be used otherwise when the classical methods cannot.

The following statistical texts will enable readers to follow up this brief survey of the field. Two lucid and entertaining introductory texts are by Moroney (1956) and Campbell (1967). Yule and Kendall's (1950) classic covers a broad field in detail but still at an introductory level, while Siegal (1956) can be recommended for distribution-free statistics. Snedecor and Cochran (1967) gives thorough treatment, with special emphasis on agricultural problems. King (1969) writes with reference to geography; R. L. Miller and Kahn (1962) and Krumbein and Graybill (1965) do the same for geology. Standard works on sampling are by Cochran (1963), Sampford (1962) and Yates (1960).

Applications of statistics in terrain evaluation can be found in the work of Beckett and Webster (1965 b, c, d), Webster and Beckett (1964, 1968, 1970), Perrin and Mitchell (1970), Brink *et al* (1968). Thornburn and co-workers (Thornburn and Larsen, 1959; Morse and Thornburn, 1961) studied soil maps made by other agencies statistically before recommending their use to engineers. Kantey and Williams (1962) appear to have been the first to test their own maps statistically to check that they could provide what the engineering user wanted. They, and Robertson and Stoner (1970) show how a careful study of terrain followed by appropriate sampling can cut costs. Beckett and Webster (1971) have reviewed measurements of soil variation mainly in the agricultural field.

19
Data management

All societies depend on the communication of information and, generally speaking, their state of advancement is a direct consequence of, and is measurable in terms of, the total rate and volume of internal information flow, expressed as *hubits* per capita per year (Meier, 1965). Hubits are units of communicated information. In most societies their number is very large. In advanced societies, they can be counted by sample surveys of the frequency with which certain critical terms are used by the press, radio, or other mass media, multiplied by the number of people who receive them. In poorer societies, their number is smaller and must be estimated from descriptions of public life and social interactions. The theory that the rate of economic development is due to the rate of information flow is called the *information theory of development*. Meier (1965) illustrates it by comparing per capita money incomes and information flow in large cities in the poorest societies (e.g. Ethiopia, Indonesia etc.) with those in completely modernised urbanised literate societies. Residents of the former had incomes estimated (in 1965) at US $40 and appeared to transmit 100 000–1 000 000 hubits per capita per year, while residents of the latter had incomes of US $1000–3000 and transmited in excess of 100 million hubits. The ratio between the incomes was at least 1:25 but not exceeding 1:75. The corresponding ratio for information flow was more than 1:500. The gap between societies in information transmission was therefore *six or seven times* greater than that between median incomes, wide as this was. Furthermore, this gap appears in the earliest stages of the development process. It is, therefore, reasonable to conclude that it is a precondition of economic progress.

Whatever weight is given to this theory, its existence underlines the importance of the acquisition and management of information about all aspects of landscape in the development process. We must, therefore, now turn to a consideration of the techniques currently employed in data storage and manipulation.

There are many organisations which store 'geographic' or 'location specific' information. They cover a wide range of interests, but are almost invariably national or larger in scale, confined to the most advanced countries, and concentrated in the fields of population census, topographic mapping and weather recording. Terrain, as distinct from topographic, factors are specifically included only by the Canadian Geographical Information system. There are broadly three directions in which geographic data storage capability can be measured, and the value of any system must be judged by the extent to which all three are satisfied. Tomlinson (1971) has expressed it in the form of a diagram, the three axes of which can be regarded as separate capability indices. Fig. 19.1 shows these indices in terms of terrain information.

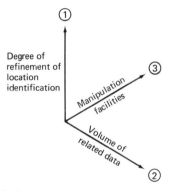

19.1 Capability indices of a data storage system

Index I represents the fineness of subdivision of the landscape and the degree of precision of the data storage system in indicating geographical location. This ranges from crude grid squares through natural units of increasing detail and refinement to a dense network of parametrically defined points, as the ultimate. Index 2 shows the amount of scientific and practical data available for each location. It can range from a minimum of one item up into the thousands. Examples of organisations with large amounts of data for each point are the US and Swedish census bureaus, and the World Wide Weather System, the latter having several thousand sequential climatic values for each station. Index 3, the manipulation facilities, is the capability of the store to handle its data. This increases from simple retrieval through a sequence of operations of increasing complexity and sophistication: summaries, scale changes, selective searches, measurements, generation of new data, automatic contouring, overlaying and comparison of data from more than one source, and ultimately the automatic monitoring of environmental changes revealed by the data. The final step is the automatic translation of the information thus obtained into decision-making and controlling activities, so that, for instance, data about pollution from a given locality can lead quickly to ameliorative action.

These operations, especially when carried out by machines, depend on a rigidly formalised recording of data. They can handle vast amounts when this is achieved. The limiting factor then becomes the sheer mechanics of storage. Each piece of information requires so many IBM or other cards or inches of microfilm or magnetic tape. Cards soon fill a large space and must be thrown out. Microfilm or tape requires less space and can be handled more rapidly, but access time becomes longer as the store grows. The US Army data storage system, for example, has only a few thousand sample points scattered round the world, but already contemplates subordinate local stations so that access to data recorded on magnetic tape can be gained within reasonable time limits (Grabau, quoted by Grant, 1968).

Most banks of terrain data are still, however, at a low level of sophistication. This is largely because of the relatively small amount of detailed location-specific ground information which is currently available for storage and the resulting limitation on the sophistication of storage techniques which is justified. This makes it especially important to consider the basic elements of data storage and the simpler techniques which are currently practicable or in use.

The elements of a data storage system for terrain

The storage of terrain information is in practice based on grid squares represented by quadrats on maps or whole aerial photographs, landscape units, or geographic coordinates which can be regarded as representative of the areas which surround them. The use of grid squares is simple and straightforward, but is complicated by the variety and the complexity of the terrain which can be included and the resulting difficulty of treating it as a unit for the storage of information. The use of landscape units is somewhat more precise and has the advantage of allowing predictions to be made between analogous areas. This is valuable in relatively unknown areas about which quick information is required. Parametrically defined terrain envelopes are a variant of the same method, but are less satisfactory in being harder to define and extrapolate. The use of grid coordinates gives the maximum precision and suitability to computer manipulation, and also allows physiographic units to be recognised if the requisite data are available. Land attributes defined in terms of x and y coordinates can be fed into a computer which can then use them to give print-outs showing terrain boundaries or to recognise analogous areas elsewhere. It should be emphasised, however, that such recognition is only possible if the data are already in the store. If not, the grid coordinate method can only provide spot information, and extrapolations must be made with the aid of existing maps and aerial photographs.

The choice of any one of the locating criteria: quadrats, land units, or grid coordinates as a basis need not be regarded as excluding the others. Storage and data manipulation techniques worked out for one may be applicable to any or all, and it is desirable, where possible, to combine their advantages in a single system. They are considered in order below.

Data storage by grid squares

Grid squares are normally only used where data are relatively sparse per unit area. The US Army Headquarters Quartermaster Research and Engineering Command use one degree quadrangles of latitude and longitude for storing numerical climatic data (Anstey, 1960).

On a finer scale, Childs (1967) has suggested relating data about natural landscape and buildings in Britain to *cubic-micro-regions*. These are based on the calculation that the biosphere extends approximately to 4 km in altitude and depth below sea level and that a surface area of 8 km square forms a useful packet. This packet is subsequently divided into 2 km square *grid units*, and these into 1 km square *component units*. The method remains only a suggestion and does not, as yet, form the basis of a data store.

The aerial photograph can be regarded as a type of grid square. Its use as a vehicle for data storage has been developed by the French Journal *Photo-Interpretation* (Editions Technip, since 1961), whose object is to disseminate specialized interpretations of aerial photographs of as many parts of the world as possible.

Each bi-monthly issue of this journal consists of about half a dozen portfolios of such photographs annotated according to a standard legend. Each portfolio interprets a single aerial photograph on four interrelated sheets. Two of these reproduce the photograph in stereo, one is transparent interpretive overlay, and one a verbal description, as illustrated on Fig. 19.2. A large number of these portfolios have now been published for which cumulative indices called *Tables analytiques* and distribution maps called *Cartes de répartition des photographies publiées* are available.

Both of the sheets with aerial photographs contain reference information and tabular keys. The reference information includes date and serial number, a letter key for the country, a note of the photograph's serial number, date, and scale, the focal length of the camera, the name of the organisation which took it, and a reference to the best available topographic map of the area represented.

The tabular key has 9 rows and 10 columns. Each row relates to a specific type of information. On the sheet containing the single photograph, the rows of the table, in order from top to bottom, relate broadly to the following type of topic.

1. Historical period of settlement in the landscape
2. Types of building or earthwork in the landscape
3. Types of non-agricultural exploitation of the landscape
4. Types of agricultural exploitation of the landscape

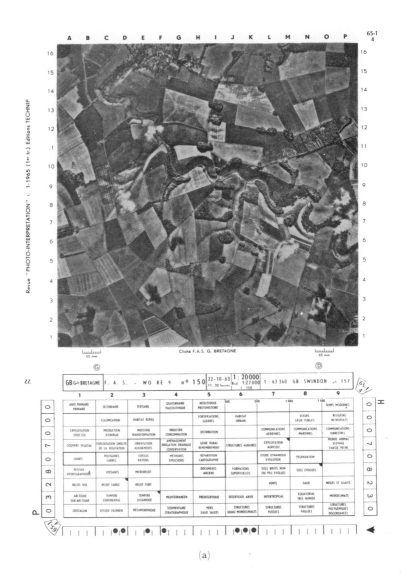

(a)

19.2 The Geotechnip method for information storage on terrain: Photo Interpretation portfolio No. 65-1 No. 4 January–February, 1965, showing the soils, physiography, and cultural pattern of part of the English Cotswolds

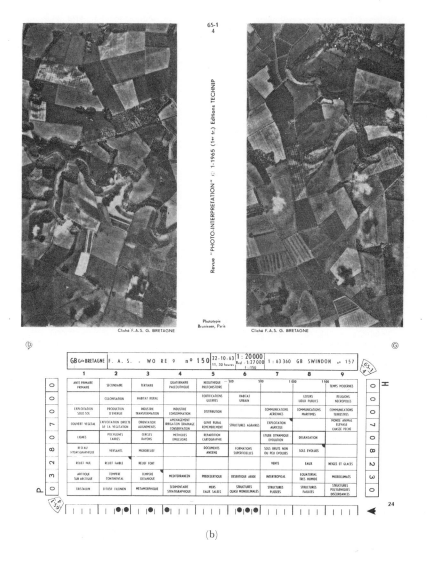

19.2 (*a*) Photograph   (*c*) Transparent interpretive overlay
    (*b*) Stereo pair   (*d*) Verbal description. (R. Webster)

GRANDE-BRETAGNE

MISSION FAS WO RE 9   Cliché n° 150 du 22/10/1963 (11 h 30) — Échelle 1 : 20 000 (réduction 1 : 27 000 environ)
focale 150 mm   Carte 1 : 63 360 (1 inch/mile)   SWINDON n° 157

THE SOILS, PHYSIOGRAPHY AND CULTURAL PATTERN OF PART OF THE ENGLISH COTSWOLDS

These photographs show a small portion of the English Cotswolds region around Eastleach, (1° 40'W, 51° 45'N), Gloucestershire. The present climate is typical of South Central England. Mean annual rainfall is approximately 700 mm.

Physiography.

The country rock is Middle Jurassic Limestone, mainly Great Oolite, which is thinly bedded, close jointed and, particularly near the surface, finely fragmented or rubbly. The dip is very gentle, about 1° towards the South-East.

The dominant landscape feature is the extensive plateau about 120 m above sea level, gently sloping and more or less parallel to the dip of the strata. This is dissected to between 30 and 40 m by the sinuous valley of the River Leach, which for most of its course is usually dry. Several smaller valleys, which begin as gentle depressions in the plateau and are now permanently dry, join the main valley.

The landscape took its present form during the Pleistocene period in a periglacial regime. The River Leach must then have been a large river probably fed by melting snow, by extremely heavy rains or a combination of both, incising its meanders into the limestone strata.

Soils.

The valley is floored by alluvial gravel of this period with a covering of finer material more recently deposited. These later became filled to a depth of one to several meters with solifluction material, soil and limestone rubble (locally known as « head ») derived from the surrounding land.

The soils on the plateau are medium to heavy textured, well drained, brown calcareous soils with much limestone rubble below 50 cm. On the steep slopes the soils are shallower, tending towards rendzinas in character, but variable owing to the varying nature of the strata on which they lie. The permanently dry valley bottoms exhibit deep well-drained brown calcareous soils on head. On the main valley floor ground water gleis are associated with the permanent stream and there are also isolated occurrences elsewhere. However, usually where stream flow is restricted to a few months during winter and spring the valley floor soils show little or no glei morphology and are moderately deep brown calcareous soils.

Cultural pattern.

From the end of mediaeval times until recently the area was used largely for sheep grazing. Indeed the Cotswolds were famous for their sheep. Recent advances in plant nutrition have allowed the land use to change to more profitable cereal growing as the main enterprise. But this change has not led to any appreciable changes in field patterns : the large regular fields of the sheep graziers suit the present day management for cereals on the plateau. The dissection slopes are usually too steep for cultivation and these remain either in pasture, or have been planted or allowed to revert to woodland. Similarly the main valley floor is scarcely worth cultivating because of its narrow tortuous configuration, even where the soils are well drained. It too is predominantly pasture.

Incidentally the Cotswold region was settled by the Romans early in the first millenium A.D. A Roman road, Akeman Street traverses the area ; its line is clearly evident on the photograph.

«British crown copyright reserved. Published by permission of the Controller, Her Britannic Majesty's Stationery Office »

R. WEBSTER
Department of Agriculture
University of Oxford

(c)

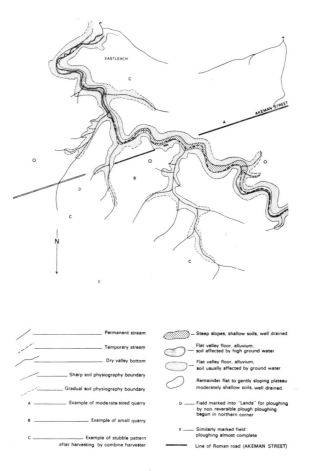

EASTLEACH

AKEMAN STREET

N

d

Permanent stream		— Steep slopes, shallow soils, well drained
Temporary stream		— Flat valley floor, alluvium; soil affected by high ground water
Dry valley bottom		— Flat valley floor, alluvium; soil usually affected by ground water
Sharp soil physiography boundary		Remainder flat to gently sloping plateau moderately shallow soils, well drained.
Gradual soil physiography boundary		
A — Example of moderate sized quarry		D — Field marked into "Lands" for ploughing by non reversible plough ploughing begun in northern corner
B — Example of small quarry		E — Similarly marked field: ploughing almost complete
C — Example of stubble pattern after harvesting by combine harvester		Line of Roman road (AKEMAN STREET)

5. Patterns visible on the photo image
6. Slopes, soils, and microrelief
7. Relief and climatic accidents
8. Climatic zone
9. Geological structure and lithology

The columns are numbered from left to right. Wherever one of these relates to the information on the photograph its number is marked in a code at the end of the row. This is shown, for instance, by the number 7 at the end of the fourth row on Fig. 19.2a, indicating that the photograph illustrates *Exploitation agricole*. Other rows are similarly marked.

It will be seen that on the other sheet containing the stereo-pairs (Fig. 19.2b), the table is inverted so that data which refer mainly to physical features appear at the top, and those which refer mainly to human features at the bottom. This ensures that the sets of cards can be filled so that either is in front depending on the type of criteria of main interest to the user.

Portfolios can be stacked sequentially, e.g. alphabetically by countries, or numerically e.g. by grid coordinates. They can be sorted according to subject matter either manually or mechanically. We have noted that each card of photographs has a numerical code. In Fig. 19.2 it is 000708230. This code is reproduced on a linear scale, with digits represented by a four-hole code, so that it can be punched out on the black spots if required. If the card with the two half-photographs is used, the spots are at the top and can either be tabbed so that the number stands up, or punched out to the edge for needle sorting.

Data storage by landscape units

Data can be stored by either physiographically or parametrically defined landscape units.

McNeil (1967) has described the basis of a terrain data store, conceived as part of the MEXE terrain evaluation programme, and the following outline is based mainly on his work. The store consists of four essential operations: input, indexing, storage, and output.

Input involves selection, reduction, and editing of incoming information in the form of literature, maps, and aerial photographs. Each item of information receives a reference number and is reduced to a standard format, called an item card, an example of which is given on Fig. 19.3. This is then photographed down on to microfilm for fast access store (see below). This will involve quartering large sheets such as maps and charts. CSIRO (Grant and Lodwick, 1968) suggest an automatic system for facilitating the transfer of field data into permanent form which might be applicable to a store of this type. Punched cards are used to record information in the field during surveys. These are then fed into a computer which prints the information out on to forms.

19.3 An item card (Design: P. H. T. Beckett)

Such forms could be adapted to suit the standard format required for the fast access store. Less important and bibliographic material is placed into a slow access store.

Indexing is designed to allow for quick random access to stored material. The simplest approach is known as the *uniterm* system using optical coordinate matching. This works as follows: each item of information is analysed for its contents and these are reduced to a number of single terms called *descriptors*. In general, these descriptors are of four types: land units, including abstracts and local forms of both land systems and facets, grid references, information about the form of items, i.e. map, photograph, table etc., and information about the content of items i.e. trafficability for jeeps, fertility for cotton, etc.

To take an example, an item of information might be that apples grow well on calcareous malmstone soils on the Lower Greensand dipslope in northern Berkshire. Now apples, calcareous malmstone soils, the sandstone dipslope facet abstract, the specific Berkshire facet local form, and the grid coordinates of the observation would each be regarded as descriptors. For each of these descriptors a *feature card* is prepared. The simplest form of feature card is an array of numbered punchable spaces. Examples of commercially available types with 1000 spaces and with 10 000 spaces are shown on Fig. 19.4. When an item relates to a certain descriptor, its number is punched on the relevant feature card. In the example quoted, the same numbered space would be punched out on the feature cards for apples, calcareous malmstone soils, sandstone dipslopes etc.

To find an item relating to two or more descriptors, one superimposes the feature cards on a light table. The places where the light shows through identify the numbers of the relevant item or items.

Feature card retrieval can be operated manually or mechanically. If there are relatively few of them, they can be indexed and placed in alphabetical order. When their number exceeds a few hundred, it is simpler to sort them with needles inserted into sorting holes around the edge. These holes are punched according to a numerical combination whereby a maximum of two holes punched out of a 'field' of four (representing respectively the numbers 1, 2, 4, and 7) can represent any digit (Casey *et al.*, 1958). When needles are inserted in a specified combination of positions, the desired cards drop into a tray. The operation of the needles can be mechanised by the use of a pneumatic actuator and valve system.

As has been indicated previously, the data store itself in which the items, as distinct from the feature cards, are kept, consists of two parts, a fast access store and a slow access store. The fast access store consists essentially of a microfilm data bank with retrieval, viewing, and printing facilities, of the sort that is now available commercially e.g. the Caps Micro-Printer made by Caps Equipment Ltd, of Colindale, London. After processing on to the standard format and indexing, all items are

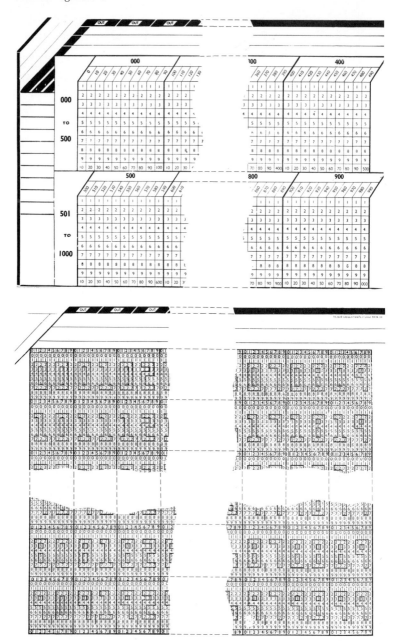

19.4 Examples of commercially available feature cards
 (*a*) with 1000 punchable spaces (Brisch-Vistem pattern)
 (*b*) with 10,000 punchable spaces (Carter-Parratt pattern)

then photographed down to occupy about one square inch of microfilm. The original material can then be destroyed or placed in the slow access store.

Updating can be done either by blanketing the old information and giving the new the same number, or by simply adding the new information as an additional item and punching out the number of the old on a coloured transparent overlay. When search is made, the overlay can be automatically inserted. Superseded items then appear in the viewer in white light while those not superseded still show the colour of the overlay.

The slow access store is the repository of all original material unsuitable for reduction or for which there is limited demand. It includes original books, articles, maps, and aerial photographs. Where the information is less important or impossible or where space must be economised, the slow access store need only contain a bibliographic reference and a note about where the work in question can be found.

The output from the store must be in the form required by the user. Simple print-outs of raw data are not adequate, and a certain amount of preparation and editing is needed. MEXE have suggested that output presentations take one of four forms: general briefs, maps, annotated aerial photographs, and specific briefs. The first three are generally provided for all users; the fourth is tailored to individual needs.

The general brief about an area is accompanied by a large scale topographic map, a land system map showing the area in question at a scale of about 1:1M, and a representative stereo-pair of aerial photographs of the land system annotated to show the facets. The brief itself contains reference information on the broad climate, geology, vegetation, and soils of the area, and illustrative block diagram showing the facets, and a table showing their main characteristics.

Specific briefs give more detailed information of interest to the specialist user. The normal form of presentation is tabular and shows the applicability of each facet for particular purposes such as 'going', airfield siting, or as a source of construction materials or water. This table is accompanied by a transparent overlay to the annotated aerial photographs showing facets and their value for the particular land use in question according to a simple legend. For instance, a specific brief for a water engineer would show the hydrological properties of the facets, while a specific brief for a fruit grower would evaluate them in terms of orchard potential.

In order to prepare the briefs for unknown areas, it is desirable to obtain information from analogous known areas. Climatic and topographic data from the unknown area is used to find a similar land system or facet abstract in the fast access store. This will indicate the local forms which appear most nearly analogous. A further scan will discard all but the most suitable. Greater caution must be exercised as

169

the closeness of the analogy and the amount of information available decrease.

A method of data storage by small composite parametric units which seek to combine the advantages of the landscape approach with those of parametric definition of land units has been explored by the Canadian Army (Parry *et al.*, 1968). Punched cards are made out for each envelope showing the parameters used to define it. These cards can then be scanned by the computer to find units which have any specified attribute or combinations of attributes.

When the attributes of a unit meet specified conditions, its number is stored in an array. Once the entire data bank has been scanned and the array formed, the sub-programme to print a map of the area with the required attribution is called.

Computerized methods of dealing with numerical data relating to grid coordinates

Some data processing and storage methods are indexed by grid co-ordinates. These have the advantage that data can be quickly trans-ferred from maps by scanning them with electronic position increasing devices, and feeding the information automatically into storage in digital form. The Headquarters Quartermaster Research and Engin-eering Command, previously mentioned, have evolved a system (Anstey, 1960) for plotting climatic data, averaged within one degree quadrangles of latitude and longitude and covering seventeen major land regions of the earth, on to maps at 1:5M. Each map normally only shows one item of information, e.g. July temperatures, January rainfall, yearly humidity, etc. The data are transferred on to punched cards according to their grid coordinates. Supplementary cards are used for quadrangles which have large internal variations (e.g. temperature means greater than 5°C, rainfall means greater than 2 in, windspeeds greater than 5 miles per hour, relative humidity values greater than 10 per cent).

A similar but far more extensive system has been devised by the Rural Development Branch of the Canadian Department of Forestry and Rural Development to manage the data acquired from the national Land Inventory. This is part of the general Canadian Geographical Information System and includes data on present and potential land uses (Tomlinson, 1968). The system entered routine use in 1968 and is in two parts: a data bank, and a set of procedures for manipulating the data once there. The bank can accept maps, of which there will be up to 30 000 for the country, and items of data related to grid references, which are entered on to magnetic tape. It is possible to measure any data in the bank relating to map areas or line lengths or to count point frequencies. Data within any given boundary can be retrieved and different types of data relating to the area within the boundary can be compared. Locations can be found with any specified characteristics. Also included is a capability for adding a reliability factor to all information and for continued modification and updating. Finally, alphabetical and numerical data can be printed out or plotted graphically. The contribution of the system is especially new in its inclusion of techniques relating to the compact storage of boundary data and the rapid comparison of one map with another. Its value necessarily depends, however, on the amount of numerical point data which can be fed into the system, and it does not easily allow for the need to use aerial photographs to recognise analogies between known and unknown areas.

The Oxford system of automatic cartography (Mott, 1967) is similar to this. A prototype instrument for producing maps automatically from data banks of individual libraries was produced by Dr Boyle of Messrs Dobbie McKinnes of Glasgow with advice and assistance from Dr David Bickmore of the Experimental Cartography Unit of the Royal College of Art. The principle is that each 'library' contains information on one type of topic. The suggested library titles were 'framework' (i.e. the basic map information), hydrology, surface cover, geology and soil, climate, buildings and built-up areas, temperatures, movement or flows, and anthropology. All information is kept on magnetic tape and the libraries are built up into a data bank. When a map is needed, the required information is fed into a computer, which assimilates it and passes it to the scanning arm of a coordinatograph which in turn draws or scribes first the basic map and then the required information. Alternatively, the information need not be printed out but may be transmitted telegraphically any distance to an intelligence subcentre. A scheme is currently in progress for making an experimental total survey of the 1:63 360 Abingdon sheet of the Geological Survey, using all possible sources of input, including satellite photography, and leading to an automatic mapped output of any data required.

Looking to the future, Dr Bickmore has said: 'We glimpse ahead the prospect of large centralised data banks assimilating geographical information on tape; modifying it, or generalising it through computers; and finally retrieving it in a variety of scales or forms ranging from the T.V. screen to the printed map' (Mott, 1967).

Means are now available for using computers to print out three-dimensional views of grid-numerical data, e.g. of heights, by methods which are known as 'geographics'. These can either be in the form of perspective drawings, or vertical or oblique images with altitudes shown by hill shading calculated in accordance with a light source of given direction and elevation (Sprunt, 1970).

Part Four

The Future

20
Prospects

At the end of this book it is desirable to survey the field of terrain evaluation as a whole in an attempt to determine where it is going, and more speculatively to suggest the direction it ought to take.

It hardly needs repeating that growing world population is leading to increasing pressure on land. Agriculture, horticulture, forestry and grazing are tending to become more selective in their choice of land and more intensive in their use of it. In many areas they are giving way to urbanisation. This trend has led to an increased concern for the conservation and improvement of the environment. There are a number of reasons for this. The higher standards of living and increased leisure of urban life have contributed. The transport revolution, particularly in the advanced countries, has made it possible to tolerate longer daily journeys to work and has allowed a general expansion of space standards. This has in turn led to an increasing variety and an accelerating geographical spread of residential and recreational land uses, bringing home to city dwellers the extent to which thoughtless exploitation of natural resources and pollution can threaten the quality of the environment. Land previously regarded as natural wilderness has suddenly acquired a considerable economic importance which is threatened by the proximity of ugliness. Where land suited to open air recreation is in short supply if it becomes used more intensively.

Associated with these developments has been the rapid spread and acceleration, especially in the past two decades, of media for the communication of information. This is especially notable in the more rapid and efficient methods of printing, copying and photography, and in the expansion of air transport, television, and the use of communications satellites such as Telstar.

We have seen that volume of information flow is an important index of, and indeed a cause of, development. One of the consequences of its increase has been to stimulate public awareness of the need for careful land conservation and development both on a national and a local scale.

For much of the world there is still only a very sparse amount of information about the land, and this is especially true of vast areas of Africa, Asia, Latin America and the polar regions; this fact must not be obscured by the significant advances that have taken place. The gap in terrain intelligence between the developed and developing parts of the world is widening.

Nevertheless, the growth of information has everywhere been relatively rapid in recent years, so that the stock is impressive when old and new sources are considered together. Libraries of books and maps have multiplied, expanded, and become accessible to more people. Aerial photographs have become generally available for much of the world, and data from other types of remote senser, carried both in aircraft and satellites, are becoming available in ever increasing quantities.

So fast has been the increase and so large is the amount of data currently in existence, that it has led to another problem: that of data management. We are no longer able adequately to store, collate, retrieve or communicate it, far less act on it. In an increasing number of areas data management has replaced data acquisition as the main bottleneck in resources evaluation and this problem is becoming more general.

These two problems are compounded by an older one: that of fragmentation of effort. The western nations operate as isolated units and little machinery exists for coordinating their resources for the task of gathering and managing worldwide terrain intelligence. They also all have formidable internal barriers which stultify or prevent the development of effective data banking systems. Commercial firms, in order to compete, are secretive. The multiplicity of government organisations, public corporations, and research institutions concerned with land use in all its aspects is much departmentalised, and they operate to a large extent in mutual isolation.

These are, however, certain general ways which, even in the absence of solutions to organisational and administrative problems, could lead in the direction of the goal of an improved general system of worldwide terrain intelligence. These relate to the widening of the approach to terrain classification and systematisation of the evaluation process.

Although the general validity of the landscape approach has been tested in temperate and arid areas, it still requires verification, especially in the polar, tundra, equatorial and tropical savanna zones. It is necessary, too, to gain a more complete cover of the world with terrain classification than exists at the moment, especially of areas which are unlikely to have their geology or soils mapped in detail. Maps showing land systems or geomorphological appreciations in similar detail today only cover parts of most of the advanced countries and a few other areas such as Uganda, Swaziland, New Guinea, Nigeria and Argentina. By far the greater part of the earth's land surface is as yet unevaluated at any scale.

Certain improvements are also needed in the current approach to the classification of terrain. First, definitions where possible need to be made more quantitative. This is especially important because of the need for the statistical analysis and computer manipulation of data. Secondly, consideration of the practical aspects of terrain should be expanded to cover the interests of a wider range of uses and academic disciplines. The emphasis hitherto has been mainly on engineering and agricultural aspects, but it is necessary also to include the microclimatic view of terrain with its emphasis on altitude, exposure and aspect and the hydrological emphasis on river catchments and terrain factors which determine run off, infiltration, and evaporation. Even more important in some areas are the visual and aesthetic aspects of landscape which affect urban and recreational development. These demand the inclusion of boundaries of the isovist or visual watershed type in the subdivision of landscape, with the aesthetic quality of the areas they include assessed in terms of quantitative visual criteria and also perhaps, in these days of pollution, according to olfactory and audial criteria as well.

This 'horizontal' widening and integration across the interdisciplinary boundaries of science concerned with land is known as *integrated survey*. There should also be a 'vertical' lengthening and integration of the operation from data acquisition to output. The aim should be something of the type known as *total survey*. In essence, this is the complete systematisation of all natural environmental resource information from initial survey up to the point at which it can be used in making planning decisions. It thus involves data acquisition, storage, manipulation, and presentation.

These stages are today handled at varying levels of sophistication. The methods of data acquisition: library study, interpretation of aerial photographs and other forms of remote sensing, field survey, laboratory and statistical analyses are fairly well established. The most serious problem is in feeding the results, which can be at any degree of detail, locational specificity, or abstraction, into a single store.

Once in the store, information must generally be held as point data on cards or magnetic tape. When a requirement is known, the information is retrieved and assembled in a form suitable for that requirement. Computer manipulation has today become a developed art. The trends are, in general, towards:

1. an increasing degree of exactness in the specification of the location to which data refers;

2. an increasing number of items of data, for each location, and an improving capability for handling them; and

3. an increasing complexity and sophistication of the calculations carried out on the stored data.

Output is by computer print-out. Considerable improvement is currently required in converting such print-outs into a form where they can be used in decision making.

Development towards total survey is already proceeding in a number of countries, but so far it is mainly experimental and covers relatively small areas only.

Looking further into the future, the ultimate objective is a more complete human understanding of, and control over, the environment. Land resource surveys prepare for planning decisions, but must become an input into the next step in the direction of such decisions – operational analysis, whose enormous potentiality has been emphasised by Grabau (Grant, 1968). This is essentially the analysis of human operations in a spatial context. As these all take place in a terrain environment, they cannot be characterised or their course predicted unless, in some prior form, this environment has also been specified. The provision of information by which this can be done is the ultimate objective of terrain evaluation.

Appendix A
Example of a method of sampling the terrain over a large area: the hot deserts of the world

Arid land shows a wide range of geological and physiographic conditions whose character and distribution have been much less studied than have those of more settled regions. The total area is so large that it is virtually impossible to study its surface conditions as a whole. A method of obtaining a representative sample of physiography is therefore generally desirable. Such a sample would make it possible to estimate the character and relative proportions of the surface covered by different landform types. This would be of assistance to many specialists concerned with terrain: to agriculturists attempting to estimate soil and water resources, to designers of cross-country vehicles, to engineers calculating requirements of local materials for engineering projects over large areas, and to anyone seeking physiographic analogies between one part of the world and another.

The most recent and comprehensive definition and classification of the arid areas of the world based directly on climatic criteria is that by P. Meigs for Unesco in 1957. It includes all areas with *aridity indices* (Thornthwaite, 1948) between minus 20 and a theoretical maximum of minus 60. Under this scheme Joly (1957) has calculated that arid lands total approximately 48 M km^2 of which 19 M km^2 are semi-arid (indices of minus 20 to minus 40), and 29 M km^2 are arid (indices below minus 40). Of the latter category 7 M km^2 are separated into a special extremely arid class if twelve consecutive months have been recorded without rainfall and there is no clear seasonal pattern.

They are distributed by continents as shown in Table A.1.

The sampling scheme is outlined below. It does not, however, quite cover the total area given on the table. The colder deserts of North America and Eurasia (i.e. Soviet and Chinese Central Asia) were excluded because of the belief that they would have appreciably different ecosystems which would give them a different physiography from that found in hotter arid areas. The remaining 24·9 M km^2 were further somewhat arbitrarily reduced by excluding all Latin American deserts (2 572 000 km^2) and the small hot arid areas in South-western Madagascar and central India (23 000 and 16 000 km^2 respectively) because of limited time and resources, the unavailability of published information and difficulties of translation. The residual area thus totalled

TABLE A.1. *Distribution of world arid lands by continents (after Joly, 1957). Values are given in thousands of square kilometres*

REGION	WITH NO MONTH AVERAGING BELOW 0°C	WITH AT LEAST ONE MONTH BELOW 0°C
Americas		
United States and Canada	682	437
South and Central America	2 572	0
Africa		
Northern hemisphere	10 807	0
Southern hemisphere	1 022	0
Madagascar	23	0
Eurasia	5 870	3 650
Australasia	3 929	0
Total	24 905	4 087

22 294 000 km² and included, broadly speaking, the Sahara, the Kalahari, south-western USA, and the Australian and Middle Eastern deserts. It is shown on Map A.1.

The sampling scheme was devised as part of the MEXE-Cambridge desert terrain evaluation project and is described by Perrin and Mitchell (1970). Briefly, physiographic units were outlined on the basis of all available geological and topographic information on 1:4M base maps. The resulting total of 238 regions are shown on maps A.2 to A.10 and listed in Table A.2, together with a representative 1° square of latitude and longitude located near to the centre of each and regarded as being representative of it. These regions were classified into the 57 lithomorphological types given in Table A.2. It was assumed that a 1° square from a single region in each of the 57 groups would represent an adequate sample of the physiography of the arid zone. Assuming a median latitude of 25°, a square of this type represents approximately 11 000 km², so that the method represents the 22 M km² of the arid areas by 57 × 11 000, or about 600 000 km², i.e. about a 2·5 per cent sample.

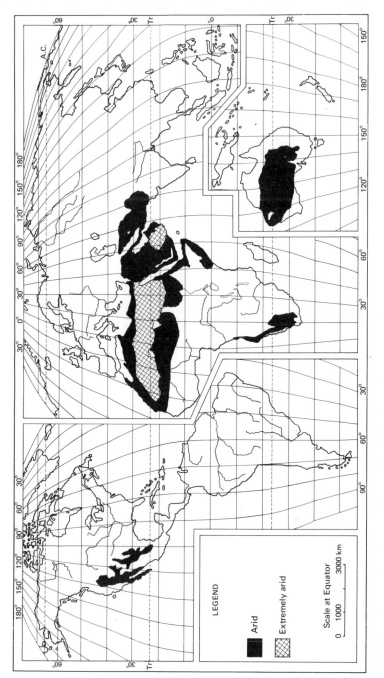

A.1 Arid and extremely arid areas of the world used for the terrain sample

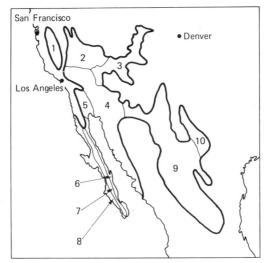

A.2 Physiographic regions of arid North America (Source: R. M. S. Perrin and C. W. Mitchell, 1970, p. 224)

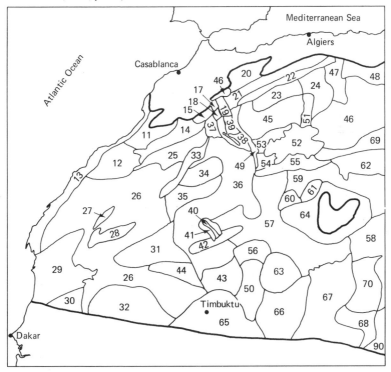

A.3. Physiographic regions of North-west Africa, Source: RM.S. Perrin and C. W. Mitchell, 1970, p. 225)

181

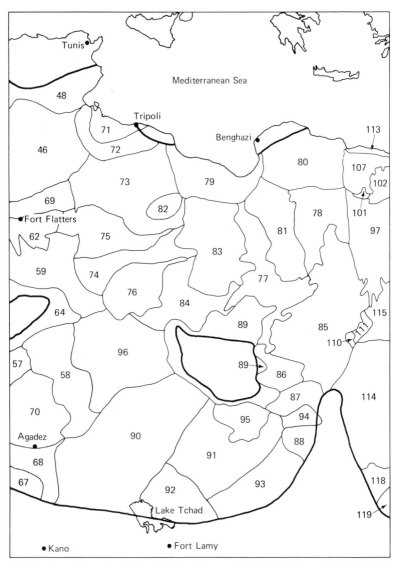

A.4 Physiographic regions of North Africa (Source: R. M. S. Perrin and C. W. Mitchell, 1970, p. 226)

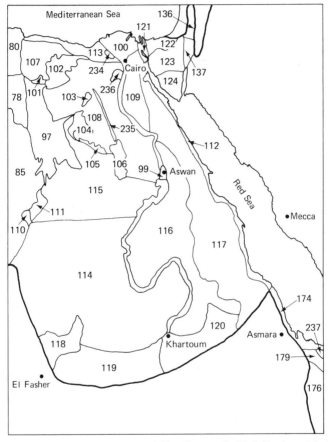

A.5 Physiographic regions of North-east Africa (Source: R. M. S. Perrin and C. W. Mitchell, 1970, p. 227)

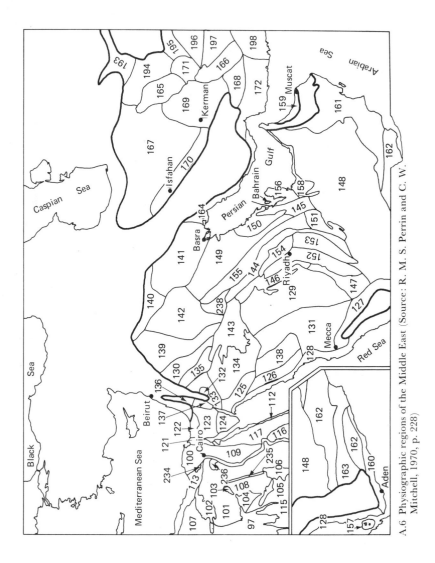

A.6 Physiographic regions of the Middle East (Source: R. M. S. Perrin and C. W. Mitchell, 1970, p. 228)

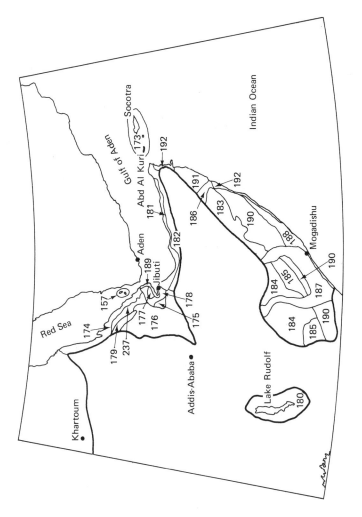

A.7 Physiographic regions of the arid lands near the Gulf of Aden (Source: R. M. S. Perrin and C. W. Mitchell, 1970, p. 229)

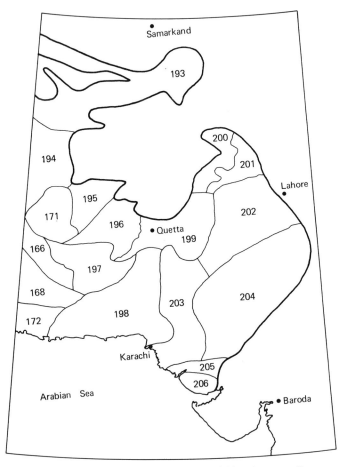

A.8 Physiographic regions of North-west India and neighbouring areas (Source: R. M. S. Perrin and C. W. Mitchell, 1970, p. 230)

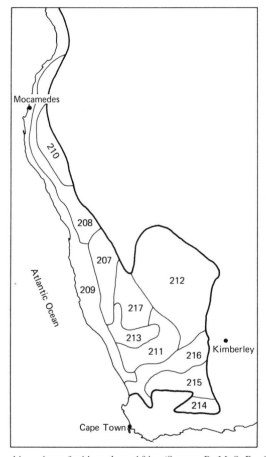

A.9 Physiographic regions of arid southern Africa (Source: R. M. S. Perrin and C. W. Mitchell, 1970, p. 231)

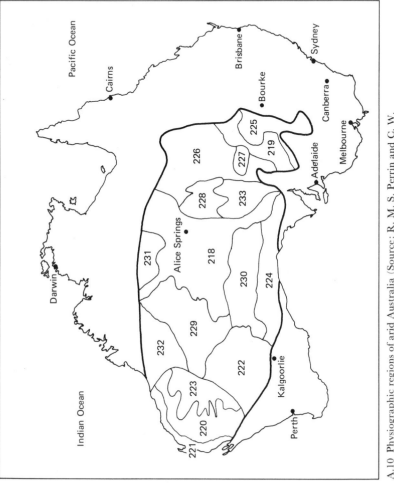

A.10 Physiographic regions of arid Australia (Source: R. M. S. Perrin and C. W. Mitchell, 1970, p. 232)

Table A.2. *Classified list of regions*

GROUP NO.	CLASS AND NAME		LIMITING COORDINATES OF REPRESENTATIVE 1° SQUARES		NO. ON MAP
	I. *Mountains*				
1	A.	Crystalline and metamorphic			
		Central ranges, Australia	23–24°S	133–134°E	218
		Mountains of S. Sinai, Egypt	28–29°S	34– 35°E	124
		Red Sea hills, Egypt and Sudan	26–27°N	33– 34°E	117
		Namaqualand Highlands, SW Africa	24–25°S	16– 17°E	207
		Adrar des Ifoghas, Algeria	19–20°N	1– 3°E	63
		Qain-Birjand Highlands, Iran	33–34°N	59– 60°E	163
		Socotran Archipelago[1]			173
		Ethiopian Coast Range			237
2	B.	Crystalline, metamorphic and volcanic mixed			
		Baja California hills, Mexico	33–34°N	109–110°W	7
		Mts of Midian, Saudi Arabia	28–29°N	35– 36°E	125
		Ahaggar foothills, Algeria	22–23°N	4– 8°E	64
		Asir-Yemen highlands, Saudi Arabia and Yemen	19–20°N	42– 44°E	127
		Ougarta mountains, Algeria	29–30°N	2– 3°W	39
3	C.	Volcanic			
		Jebel Harug, Libya	27–28°N	17– 18°E	83
		Jebel Es Soda, Libya	28–29°N	15– 16°E	82
		Mountain areas of French Somaliland	11–12°N	42– 43°E	177
		Hanich islets in Red Sea	13–14°N	42– 43°E	157
		Jebel Marra, Darfur, Sudan			118
4	D.	Sandstone			
		Jebel Uweinat – Archenu, Egypt – Sudan – Libya	21–22°N	25– 26°E	110
5	E.	Limestone and Dolomite			
		Mekran, W Pakistan	25–26°N	64– 65°E	198
		N Ranges, Baluchistan, W Pakistan	29–30°N	68– 69°E	199
		NW hills, W Pakistan	33–34°N	72– 73°E	200
		Saharan Atlas, Algeria	33–34°N	1– 2°E	21
		Jebel Nefusa – Ksour, Libya and Tunisia	31–32°N	11– 12°E	72
	II. *Hills*				
6	A.	Foothills: Limestone			
		Ogaden Fringe, Somalia – Ethiopia	7– 8°N	47– 48°E	183
		Foothills of Somalia ('Ogo')	10–11°N	45– 46°E	182
		Shebeli-Juba Plateau, Somalia–Kenya	3– 4°N	41– 42°E	184

1. Regions without representative 1° squares were not on the list originally used. They are included for the sake of completeness.

TABLE A.2. *continued*

GROUP NO.	CLASS AND NAME	LIMITING COORDINATES OF REPRESENTATIVE 1° SQUARES		NO. ON MAP
7 B.	Undulating country			
	1. Crystalline with volcanics			
	Air Massif, Niger	17–18°N	8– 9°E	70
8	2. Foliated:			
	Barrier-Grey Ranges, Australia	28–29°S	143–144°E	219
9	3. Limestone:			
	Hamad plateau, Arabia–Jordan–Iraq– Syria	32–33°N	39– 40°E	139
	Upper Tigris–Euphrates plains, Iraq	33–34°N	42– 43°E	140

III. *Piedmonts*

GROUP NO.	CLASS AND NAME	LIMITING COORDINATES OF REPRESENTATIVE 1° SQUARES		NO. ON MAP
10 A.	Erosional			
	1. Crystalline			
	NW Slopes of NW Tableland, Australia	23–24°S	115–116°E	220
	Uplands of NW SW Africa	19–20°S	13– 14°E	208
11	2. Volcanic			
	Central zone of French Somaliland	11–12°N	42– 43°E	178
12	3. Limestone			
	Persian Gulf Coastal plain, Iran	29–30°N	50– 51°E	164
13 B.	Depositional			
	1. Mainly calcareous			
	Batina coastal plain, Muscat	23–24°N	57– 58°E	159
	Djeffara coastal plain, Libya–Tunisia	32–33°N	12– 13°E	71
	Coastal strips of S Somalia			192
	Egyptian coastal plain W of Delta	31–32°N	27– 28°E	113
	Red Sea coastal plain Egypt–Sudan	26–27°N	34– 35°E	112
	S slopes of Saharan Atlas, Algeria			22
14	2. Mainly acidic			
	Aden coastal plain	12°30′– 13°30′N	44°30′– 45°30′E	160
	Coastal plain of W Australia	23–24°S	113–115°E	221
	San Joachim valley basin. California, USA	36–37°N	119–120°W	1
	Baja California Gulf coastal plain, Mexico	33–34°N	109–110°W	6
	Baja California Pacific coastal plain, Mexico	33–34°N	110–111°W	8
	Tihama coastal plain, Saudi Arabia– Yemen	19–20°N	41– 42°E	128
	Ethiopian coastal plain	15–16°N	39– 40°E	174
	Namib coastal plain, Angola, SW Africa and S Africa	19–20°S	12– 13°E	209

TABLE A.2. *continued*

GROUP NO.	CLASS AND NAME	LIMITING COORDINATES OF REPRESENTATIVE 1° SQUARES		NO. ON MAP

IV. *Plateaux*

15 A.	Slightly dissected			
	1. Crystalline or metamorphic			
	Salt Lake Division, Australia	28–29°S	120–123°E	222
	Bur Region, Kenya	1– 2°N	40– 41°E	185
	Hasa Coastal Desert Saudi Arabia	26–27°N	49– 50°30′E	150
	Eglab, Algeria	26–27°N	3– 4°W	34
	Central crystalline plateau, Saudi Arabia	24–25°N	44– 45°E	129
16	2. Volcanic 'harras'			
	Northern Harras Area, Saudi Arabia	31–32°N	38– 39°E	130
	Central Harras Area, Saudi Arabia	22–23°N	41– 42°E	131
	Plateaux of French Somaliland	11–12°N	42– 43°E	175
	Nasratabad-Taftan Highlands, Iran	28–29°N	61– 62°E	166
17	3. Limestone: with some vegetation			
	High plateau of Chotts, Algeria	33–34°N	1– 2°W	20
18	4. Limestone: bare			
	Nullarbor Plain, Australia	30–31°S	130–131°E	224
	Altoplanice, Mexico, USA	32–33°N	107–108°W	9
	Limestone plateau of SE Jordan	30–31°N	37– 38°E	133
	Hamada of Murzuch, Libya	26–27°N	12– 13°E	74
	Hamada of Tademait, Algeria	28–29°N	1– 2°E	52
	Hamada of Dra, Algeria	28–29°N	7– 8°W	14
	Hamada of Daoura, Algeria	30–31°N	3– 4°W	18
	Hamada of El Gantara, Algeria	30–31°N	3– 4°E	51
	Hamada of Bou Denib, Algeria–Morocco	32–33°N	3– 4°W	16
	Hamada of Bou Laouaiche, Morocco	31–32°N	4– 5°W	17
	Hamada of El Haricha, Mali	22–23°N	3– 4°W	40
	Hamadas of NE Spanish Sahara			12
	Hamadas of S Morocco			11
	Hamadas of Krenachich-Mahia, Mali			42
	Hamada El Homra, Libya	29–30°N	12– 13°E	73
	Interior Plateau, Baluchistan, W Pakistan	28–29°N	63– 64°E	197
	Central Plateau, Egypt	26–27°N	30– 31°E	108
	NW Plateau, Egypt	30–31°N	26– 27°E	107
19	5. Sandstone			
	South African Highveld	31–32°S	22– 23°E	215
	Cape Middleveld, South Africa			216
	Gilf Kebir, Egypt	23–24°N	26– 27°E	111
	Hamada of Guir, Algeria–Morocco	30–31°N	3– 4°W	19
	Tindouf,	27–28°N	7– 8°W	25
	Araouane,	25–26°N	3– 4°W	35
	Sandstone plateau of N Hejaz, Saudi Arabia	28–29°N	37– 38°E	34
	Mauritanian interior plateau	23–24°N	9– 10°W	26
	Jefjef plateau, Tchad–Libya	20–21°N	20– 21°E	86
	Coastal plain of S Morocco and Spanish Sahara	30–31°N	9– 10°W	13

TABLE A.2. *continued*

GROUP-NO.		CLASS AND NAME	LIMITING COORDINATES OF REPRESENTATIVE 1° SQUARES		NO. ON MAP
20	B.	Moderately dissected			
		1. Crystalline or metamorphic			
		NW tableland of Australia	23–24°S	116–118°E	223
		Cutch, India	23–24°N	69– 80°E	206
		Sol plateau, Somalia			191
21		2. Limestone			
		Al Wadiyan Area, Saudi Arabia–Iraq	32–33°N	41– 43°E	142
		Dibdibba-Hasa Plains, Saudi Arabia	29–30°N	45– 46°E	149
		Hajara Plain, Saudi Arabia	29–30°N	42– 43°E	238
22		3. Sandstone			
		Great Karroo, S Africa	32–33°S	23– 24°E	214
		Erdi plateau, Tchad–Libya	19–20°N	22– 23°E	87
		Sandstone plateau SE of Aswan			116
	C.	Much dissected			
23		1. Crystalline or foliated			
		Central Hejaz Uplands, Saudi Arabia	24–25°N	38– 39°E	126
		Jabrin plateau, Saudi Arabia	23–24°N	48– 49°E	151
		Tuweiq plateau, Saudi Arabia	25–26°N	45– 46°E	152
		Biyadh plateau, Saudi Arabia	23–24°N	47– 48°E	153
		Arma plateau, Saudi Arabia	25–26°N	46– 47°E	154
		Summan plateau, Saudi Arabia	25–26°N	48– 49°E	155
24		2. Limestone (including kem-kems)			
		Chela-Otair Highlands, SW Africa	19–20°S	14– 15°E	210
		Jol and Kathiri-Mahra Plateaux,			
		Aden	15–16°N	49– 50°E	162
		Kem-Kem, Algeria	30–31°N	4– 5°W	15
		Mzab Plateau, Algeria	32–33°N	3– 4°E	24
		Tinghert Plateau, Algeria	28–29°N	6– 7°E	69
		Central Sinai–Negev, Egypt–Israel	30–31°N	34– 35°E	123
		Maaza Plateau, Egypt	26–27°N	32– 33°E	109
		Terecht–Timetrim, Mali	21–22°N	0– 1°E	56
		Ain Sefra plateau, Algeria			23
25		3. Sandstone ('tassili')			
		Tibesti outliers, Libya	21–22°N	19– 20°E	89
		Tassili, Algeria	25–26°N	8– 9°E	59
		Ahnet, Algeria	24–25°N	2– 3°E	60
		Mouydir, Algeria	25–26°N	4– 5°E	61
		Mekran, Iran	26–27°N	58– 59°E	172
		Ennedi, Tchad	17–18°N	22– 23°E	88
		V. *Desert plains*			
	A.	With inselbergs			
		1. *Gravelly*			
26		(*a*) with crystalline inselbergs			
		Plains of E and Air Massif, Niger	18–19°N	11– 12°E	96
27		(*b*) with volcanic inselbergs			
		Farah lowlands, SW Afghanistan	33–34°N	61– 62°E	194

TABLE A.2. *continued*

GROUP NO.	CLASS AND NAME	LIMITING COORDINATES OF REPRESENTATIVE 1° SQUARES		NO. ON MAP
28	(c) with limestone inselbergs			
	Qatar and Bahrein	25–26°N	51– 52°E	156
	2. Sandy			
29	(a) with crystalline inselbergs			
	Great Basin, USA	37–28°N	117–118°W	2
	Sonoran desert, USA	32–34°N	113–116°W	4
	W plains of NSW and Queensland, Australia	28–29°S	144–145°E	225
	E Namaqualand, SW Africa			217
	Little Namaqualand–Bushmanland plain, SW Africa	29–30°S	19– 20°E	211
	S Kalahari, SW Africa	26–27°S	20– 21°E	212
	Mortcha, Tchad	16–17°N	20– 21°E	93
30	(b) with volcanic inselbergs			
	Lake Rudolf Basin, Kenya	3– 4°N	35– 36°E	180
	E lowlands of Ethiopia	12–13°N	41– 42°E	176
31	(c) with limestone inselbergs			
	Trucial Coast, Arabia	24–25°N	54– 55°E	158
	Sand and gravel plains of Oman– Dhofar	19–20°N	57– 58°E	161
	Lowlands W of Air, Niger	18–19°N	5– 6°E	67
	Guban, Somalia	10–11°N	45– 46°E	181
32	(d) with sandstone inselbergs			
	SE Desert Plains, Libya	22–23°N	22– 23°E	85
	Central N Desert of Sudan	21–22°N	30– 31°E	114
	Cent Desert Plains, Libya	25–26°N	17– 18°E	84
	S Plateau, Egypt	21–22°N	28– 29°E	115
	Borkou, Tchad	18–19°N	20– 21°E	95
	Thar Desert, W Pakistan–India	26–27°N	70– 71°E	204
	Ténéré, Algeria	22–23°N	9– 10°E	58
33	3. Argillaceous			
	Low plains, Somalia[1]			190
	Mauritanian coastal plain	19–20°N	16– 17°W	29
34	B. Without inselbergs			
	1. Gravelly			
	Serir of Calansho, Libya	27–28°N	22– 23°E	81
	Marmarica Lowlands, Libya	31–32°N	24– 25°E	80
	Tanezrouft, Algeria	23–24°N	0– 1°E	57
	Dasht-i-Margo Plain, Afghanistan	31–32°N	63– 64°E	195
	Tagama Plain, Niger	16–17°N	8– 9°E	68
	SE Plains of Mali[1]			66
35	2. Sandy[1]			
	E central lowland of Australia	24–25°S	139–141°E	226
	Sirte lowlands, Libya	30–31°N	18– 19°E	79
	Sinai coastal plain, Egypt	28–29°N	33– 34°E	122
	Djouf, Mauritania[1]			31
	Plains of Aouker and Hodh, Mali[1]			32
	Majabat Al Koubra, Mali[1]			44
	Isthmus of Suez, Egypt	30–31°N	32– 33°E	121

193

Tᴀʙʟᴇ A.2. *continued*

GROUP NO.	CLASS AND NAME	LIMITING COORDINATES OF REPRESENTATIVE 1° SQUARES		NO. ON MAP
	Kanem, Tchad	14–15°N	15– 16°E	92
	Azaouad, Mali	18–19°N	2– 3°W	43
	Manga, Niger	14–15°N	13– 14°E	90
	Registan, Afghanistan	30–31°N	65– 66°E	196
36	3. Clayey			
	Sudan clay plains			120
37	4. Lateritic			
	Brakna-Douaich plains, Mauritania	17–18°N	13– 14°W	30
	VI. *Main valleys*			
38	A. Large seasonal washes			
	Wadi Hadhramaut, Aden	15–16°N	48– 49°E	163
	Igharghar–Rhir Basin, Algeria	32–33°N	6– 7°E	47
	Tabelbala Basin, Algeria	29–30°N	3– 4°W	37
	Wadi Saoura, Algeria	29–30°N	1– 2°W	49
	Wadi Tlemsi, Mali	17–18°N	0– 1°E	50
	B. Upper eroding areas of perennial rivers			
39	1. Crystalline			
	Orange River Gorge Tract, SW Africa	28–29°S	17– 18°E	213
	Shabluka Gorge, Sudan	16–17°N	32– 33°E	98
	Aswan Cataract, Egypt	24–25°N	32– 33°E	99
40	2. Sandstone			
	Grand Canyon, USA	36–37°N	112–113°W	3
	Pecos Valley, Texas, USA	31–32°N	103–104°W	10
41	3. Limestone			
	Submontane Indus tract, W Pakistan	33–34°N	71– 72°E	201
	Oxus plains, Afghanistan	37–38°N	66– 67°E	193
	C. Middle alluvial areas			
42	1. Siliceous			
	Nile valley, Egypt and Sudan	26–27°N	31– 32°E	100
	Sind Plain, W Pakistan	27–28°N	68– 69°E	203
	Punjab Plain, W Pakistan	30–31°N	72– 73°E	202
43	2. Calcareous			
	Tigris–Euphrates Lowlands, Iraq	32–33°N	35– 46°E	141
	Nogal Valley System, Somalia	8– 9°N	49– 50°E	186
	Shebeli-Juba Lowlands, Somalia	1– 2°N	43– 44°E	187
	D. Lower Swampy courses			
44	1. Siliceous			
	Wadi Sirhan, Jordan	31–32°N	37– 38°E	135
	Niger Valley, Mali	16–17°N	2– 3°W	65
45	2. Calcareous			
	Jordan Valley, Jordan	31–32°N	35– 36°E	136

TABLE A.2. *continued*

GROUP NO.	CLASS AND NAME	LIMITING COORDINATES OF REPRESENTATIVE 1° SQUARES		NO. ON MAP
	VII. *Eolian areas : ergs*			
46	Fixed dune areas (qoz) of Kordofan, Sudan			119
	Sturt desert, Australia	28–29°S	141–143°E	227
	Simpson desert, Australia	24–25°S	136–137°E	228
	Gibson desert, Australia	24–25°S	126–127°E	229
	Gt Victorian desert, Australia	28–29°S	130–131°E	230
	NT Sand desert, Australia	20–21°S	132–133°E	231
	Gt Sandy desert, Australia	21–22°S	123–124°E	232
	Rebiana sand sea, Libya	24–25°N	21– 22°E	77
	Calansho sand sea, Libya			78
	Gt Egyptian sand sea, Egypt	26–27°N	26– 27°E	97
	Abu Muharik dunes, Egypt	26–27°N	30– 31°E	235
	Edeien Murzuch, Libya	24–25°N	13– 14°E	76
	Edeien Ubari, Libya	27–28°N	12– 13°E	75
	Gt E Erg, Algeria	30–31°N	7– 8°E	46
	Gt W Erg, Algeria	30–31°N	0– 1°E	45
	Erg Er Raoui	28–29°N	1– 2°W	38
	Erg Issaouane, Algeria			62
	Erg Chech, Algeria			36
	Erg Iguidi, Algeria			33
	Erg Hamami, Algeria			27
	Erg Makteir, Algeria			28
	Rub Al Khali, Saudi Arabia	19–20°N	50– 51°E	148
	Gt Nafud, Saudi Arabia	28–29°N	40– 41°E	143
	Nafud Dahi, Saudi Arabia	22–23°N	45– 46°E	147
	Jafura, Saudi Arabia	24–25°N	50– 51°E	145
	Inner Girdle Sand Desert, Saudi Arabia	25–26°N	44– 45°E	146
	Outer Girdle Sand Desert, Saudi Arabia or	{ 25–26°N 23–24°N	47– 48°E } 48– 49°E }	144
	Coastal consolidated dunes, Somalia	2– 3°N	45– 46°E	188
	VIII. *Large enclosed depressions*			
47 A.	Tectonic or fold			
	1. Internal drainage and salt lakes			
	Lake Eyre Basin, Australia	28–29°S	137–138°E	233
	Danakil Depression, Ethiopia	13–14°N	40– 41°E	179
	Chotts, Tunisia	33–34°N	8– 9°E	48
	Great Kavir, Iran	34–35°N	53– 54°E	167
	Jaz Murian, Iran	27–28°N	58– 59°E	168
	S Lut, Iran	30–31°N	58– 59°E	169
	Isfahan–Siran, Iran	32–33°N	52– 53°E	170
	Sistan, Iran[1]			171
	Bodelé ('pays-bas'), Tchad	16–17°N	16– 17°E	91
48	2. with external drainage			
	Salton Trough, California, USA	33–34°N	115–116°W	5

TABLE A.2. *C*

GROUP NO.		CLASS AND NAME	LIMITING COORDINATES OF REPRESENTATIVE 1° SQUARES		NO. ON MAP
		Arabah Graben, Jordan	30–31°N	35– 36°E	137
		Murdi Tchad	18–19°N	21– 22°E	94
		Al Medina Basin, Saudi Arabia	25–26°N	38– 39°E	138
	B.	Large solution and deflation hollows			
49		1. Sandstone areas			
		Touat, Algeria	27–28°N	0– 1°E	54
50		2. Limestone areas			
		Jefr depression, Jordan			132
		Kharga Oasis, Egypt	25–26°N	30– 31°E	106
		Dakhla Oasis, Egypt	25–26°N	29– 30°E	105
		Bahariya Oasis, Egypt	28–29°N	28– 29°E	103
		Farafra Oasis, Egypt	27–28°N	28– 29°E	104
		Fayum and Wadi Rayan, Egypt[1]			236
		Siwa Oasis, Egypt	29–30°N	25– 26°E	101
		Qattara Depression, Egypt	29–30°N	27– 28°E	102
		Wadi Natrun, Egypt	30–31°N	30– 31°E	234
		Tidikelt Oasis, Algeria	27–28°N	2– 3°E	55
		Gourara Oasis, Algeria	28–29°N	0– 1°E	53
		Taoudeni Oasis, Mali	22–23°N	3– 4°E	41
51	C.	Interior basins with external drainage			
		Plains of French Somaliland	11–12°N	42– 43°E	189
52		IX. *Intertidal features : marsh*			
		Rann of Cutch, India	23–24°N	68– 69°E	205
				TOTAL	236

TABLE A.3. *Key to position of regions in classified list*

NO. ON MAP	GROUP	NO. ON MAP	GROUP	NO. ON MAP	GROUP	NO. ON MAP	GROUP	NO. ON MAP	GROUP	NO. ON MAP	GROUP	NO. ON MAP	GROUP
1	14	35	19	69	24	103	50	137	48	171	47	205	52
2	29	36	46	70	7	104	50	138	48	172	25	206	20
3	40	37	38	71	13	105	50	139	9	173	1	207	1
4	29	38	46	72	5	106	50	140	9	174	14	208	10
5	48	39	2	73	18	107	18	141	43	175	16	209	14
6	14	40	18	74	18	108	18	142	21	176	30	210	24
7	2	41	50	75	46	109	24	143	46	177	3	211	29
8	14	42	18	76	46	110	4	144	46	178	11	212	29
9	18	43	35	77	46	111	19	145	46	179	47	213	29
10	40	44	35	78	46	112	13	146	46	180	30	214	39
11	18	45	46	79	35	113	13	147	46	181	31	215	22
12	18	46	46	80	34	114	32	148	46	182	6	216	19
13	19	47	38	81	34	115	32	149	21	183	6	217	19
14	18	48	47	82	3	116	22	150	15	184	6	218	29
15	24	49	38	83	3	117	1	151	23	185	15	219	1
16	18	50	38	84	32	118	3	152	23	186	43	220	8
17	18	51	18	85	32	119	46	153	23	187	43	221	10
18	18	52	18	86	19	120	36	154	23	188	46	222	14
19	19	53	50	87	22	121	35	155	23	189	51	223	15
20	17	54	49	88	25	122	35	156	28	190	33	224	20
21	5	55	50	89	25	123	24	157	3	191	20	225	18
22	13	56	24	90	35	124	1	158	31	192	13	226	29
23	24	57	34	91	47	125	2	159	13	193	41	227	35
24	24	58	32	92	35	126	23	160	14	194	27	228	46
25	19	59	25	93	29	127	2	161	31	195	34	229	46
26	19	60	25	94	48	128	14	162	24	196	35	230	46
27	46	61	25	95	32	129	15	163	38	197	18	231	46
28	46	62	46	96	26	130	16	164	12	198	5	232	46
29	33	63	1	97	46	131	16	165	1	199	5	233	46
30	37	64	2	98	39	132	50	166	16	200	5	234	47
31	35	65	44	99	39	133	18	167	47	201	41	235	50
32	35	66	34	100	42	134	19	168	47	202	42	236	46
33	46	67	31	101	50	135	44	169	47	203	42	237	50
34	15	68	34	102	50	136	45	170	47	204	32	238	21

Further reading

As can be seen from the list of references, the literature on terrain is extensive and comes from a wide variety of sources, but few books are wholly devoted to it.

There is a primary distinction between academic works on the geomorphic analysis and classification of terrain on the one hand and those on the study of its applications to practical needs on the other. Among the former must be included standard works on geomorphology and physical geography such as von Engeln's *Geomorphology* (1942), Thornbury's *Principles of Geomorphology*, and Strahler's *Introduction to Physical Geography* (1967), even though these are concerned as much with origins and processes as with the analysis and classification of surface form. Examples of some of the most fundamental contributions to terrain classification are Linton's 'The delimitation of morphological regions' (in Stamp and Wooldridge, *London Essays in Geography*, 1951), Fenneman's 'Physiographic divisions of the United States' (*Annals of the Association of American Geographers*, 1928), Savigear's 'A technique of morphological mapping' (*Annals of the Association of American Geographers*, 1965), and Verstâppen and van Zuidam's *ITC System of Geomorphological Survey* (1968).

The only general text on the applications of terrain study to practical needs in agriculture, engineering, planning, etc. is the collection of essays under the editorship of G. A. Stewart entitled *Land Evaluation* (1968). This is both authoritative and up-to-date, containing papers by many of the leaders in the field including a review of concepts by J. A. Mabbutt. It is, however, somewhat advanced for the general reader, carrying most aspects up to the present research frontiers.

Within a more restricted compass, there are a number of summaries done by various organisations in this field. The chief of these are Beckett and Webster's *A review of studies on terrain evaluation by the Oxford-Cambridge-MEXE Group* (MEXE Report 1123, 1969), Mabbutt and Stewart's 'The applications of geomorphology in resources surveys in Australia and New Guinea' (*Revue de Géomorphologie Dynamique*, 1965), and the US Army Engineer Waterways Experiment Station's *Military evaluation of geographic areas: reports on activities to 1963* (Miscellaneous Paper No. 3–610, 1963).

References

ADDOR, E. E. (1963) 'Vegetation description for military purposes', in US Army Engineer Waterways Experiment Station, *Military Evaluation of Geographic Areas, Reports on activities to April, 1963*, Miscellaneous Paper No. 3–610.

AFANASIEV, J. N. (1927) 'The classification problem in Russian soil science', *Russian Pedology* **5**. Academy of Sciences of the USSR. Leningrad; quoted in US Department of Agriculture (1960), *Soil classification: a comprehensive system*, 7th Approximation, 6.

AIR FORCE CAMBRIDGE RESEARCH LABORATORIES (US AIR FORCE): *see* Motts, Ta Liang.

AITCHISON, G. D. and GRANT, K. (1967) 'The P.U.C.E. programme of terrain description, evaluation, and interpretation for engineering purposes', *Proceedings of the Fourth Regional Conference for Africa on Soil Mechanics and Foundation Engineering*, Cape Town.

AITCHISON, G. D. and GRANT, K. (1968a) *Proposals for the application of the P.U.C.E. programme of terrain classification and evaluation to some engineering problems*. Paper No. 452T for the Symposium on Terrain Evaluation for Engineering in Aug. 1968 convened by the Division of Soil Mechanics, CSIRO, in association with the Fourth Conference of the Australian Road Research Board. Melbourne. CSIRO Division of Soil Mechanics Research Paper no. 119.

AITCHISON, G. D. and GRANT, K. (1968b) 'Terrain evaluation for engineering, in Stewart, ed. (1968), pp. 125–46.

ANONYMOUS (1965) 'Terrain profiler', *American Geological Institute*, **9**, 26; quoted in Stone and Dugundji (1965).

ANSTEY, F. (1960) *Digitised environmental data processing*, HQ Quartermaster Research and Engineering Command. US Army, Natick, Mass. Research Study Report RER–31. Project no. 7–83–01–007.

BECKETT, P. H. T. (1962) 'Punched cards for terrain intelligence', *Royal Engineers Journal*, **76**, no. 2, 185–93.

BECKETT, P. H. T. (1967) *Report of a visit to the terrain evaluation cell of the R and D Organisation of the Indian Army*, MEXE Report no. 1020.

BECKETT, P. H. T. and WEBSTER, R. (1962) *The Storage and Collation of Information on Terrain*, MEXE Report, Christchurch, Hants (Part 1 by P. H. T. Beckett, Part 2 by R. Webster).

References

BECKETT, P. H. T. and WEBSTER, R. (1965a) *A Classification System for Terrain*, MEXE Report no. 872, An Interim Report, Christchurch, Hants.

BECKETT, P. H. T. and WEBSTER, R. (1965b). *Field Trials of a Terrain Classification System: organisation and methods*, MEXE Report no. 873.

BECKETT, P. H. T. and WEBSTER, R. (1965c) *Field Trials of a Terrain Classification System: statistical procedure*. MEXE Report no. 874.

BECKETT, P. H. T. and WEBSTER, R. (1965d) *Minor Statistical Studies on Terrain Evaluation*, MEXE Report no. 877.

BECKETT, P. H. T. and WEBSTER, R. (1969) *A Review of Studies on Terrain Evaluation by the Oxford-MEXE-Cambridge group, 1960–1969*, MEXE Report no. 1123.

BECKETT, P. H. T. and WEBSTER, R. (1971) 'Soil variability: a review', *Soils and Fertilisers*, **34**, 1–15.

BERRY, B. J. L. and MARBLE, D. F. eds. (1967) *Statistical Analysis: a reader in physical geography*, Prentice-Hall.

BIBBY, J. S. and MACKNEY, D. (1969) *Land Use Capability Classification*, Soil Survey Technical Monograph no. 1. Rothamsted Experimental Station and Macaulay Institute for Soil Research.

BLACK, C. A., ed. (1965) 'Methods of soil analysis' *Agronomy*, **9**, American Society of Agronomy, Madison, Wis.

BLANEY, H. F. and CRIDDLE, W. D. (1950) *Determining Water Requirements in Irrigated Areas from Climatological and Irrigation Data*, US Department of Agriculture Soil Conservation Service.

BOURNE, R. (1931) 'Regional survey and its relation to stocktaking of the agricultural resources of the British Empire', *Oxford Forestry Memoirs*, no. 13.

BOWMAN, I. (1914) *Forest Physiography, Physiography of the US and Principal Soils in Relation to Forestry*.

BOWMAN, I. (1916) *The Andes of Southern Peru*, American Geographical Society Publication, no. 2, New York.

BRIDGES, E. M. and DOORNKAMP, J. C. (1963) 'Morphological mapping and the study of soil patterns', *Geography*, **48**, no. 2, 175–81.

BRINK, A. B. A., MABBUTT, J. A., WEBSTER, R., and BECKETT, P. H. T. (1965) *Report of the Working Group on land classification and data storage*, MEXE Report no. 940.

BRINK, A. B. A., PARTRIDGE, T. C., WEBSTER, R., and WILLIAMS, A. A. B. (1968) 'Land classification and data storage for the engineering usage of natural materials', *Proceedings of the Symposium on Terrain Evaluation for Engineering*. Australian Road Research Board 4th Conference, **4**, 1624–47.

BRITISH STANDARDS INSTITUTION (1967) *Methods of testing soils for civil engineering purposes*, British Standard 1377.

BROUGHTON, J. D. and ADDOR, E. E. (1968) *Mobility environmental research study. A quantitative method for describing terrain for ground mobility*, vol. 4. *Vegetation*. Technical Report no. 3–726, USAEWES.

BROWN, G., ed. (1961) *The X-ray Identification and Crystal Structures of Clay Minerals* Mineralogical Society Clay Minerals Group, London.

CAMPBELL, R. C. (1967) *Statistics for Biologists*, Cambridge University Press.

CANADA: DEPARTMENT OF MINES AND TECHNICAL SURVEYS (1969) *Atlas of Canada*.

CANADA LAND INVENTORY (1965). Report no. 2, *Soil Capability Classification for Agriculture*.

CARR, D. D., BECKER, R. E. and VAN LOPIK, J. R. (1963) *Terrain Quantification*, Phase 2. Playa and miscellaneous studies. Texas Instruments Inc. Science Services Division Contract no. AF19 (628)–2786, Project no. 7628, Task no. 762805, US Air Force.

CASEY, R. S., PERRY, J. W., BERRY. M. M. and KENT, A. A., eds. (1958) *Punched Cards: their applications to science and industry*. Reinhold, New York.

CHAPMAN, T. G. (1968) 'Catchment parameters for a deterministic rainfall-runoff model', in Stewart (1968), pp. 312–23.

CHILDS, D. R. (1967) 'The Anatomy of Human Environment', in Landscape Research Group Symposium (1967), pp. 43–6.

CHOW, VEN TE, ed. (1964) *Handbook of Applied Hydrology*, McGraw-Hill, New York.

CHRISTALLER, W. (1933) *Die Zentralen Orte in Suddeutschland*, Jena.

CHRISTIAN, C. S. (1958) 'The concept of land units and land systems', *Proceedings of the 9th Pacific Science Congress, 1957*, **20**, 74–81.

CHRISTIAN, C. S. and STEWART, G. A. (1964) *Methodology of Integrated Surveys* Unesco Conference on Principles and Methods of Integrating Aerial Survey Studies of Natural Resources for Potential Development, Toulouse.

CLARKE, G. R. (1957) *The study of the soil in the field*, 4th edn, Oxford, Clarendon Press.

CLOUSTON, J. B. (1967) *The Durham Motorway Landscape Study*, in Landscape Research Group Symposium (1967), pp. 11–19.

COCHRAN, W. G. (1963) *Sampling Techniques*, 2nd edn, Wiley, New York.

COLE, J. P. and KING, C. A. M. (1968) *Quantitative Geography*, J. Wiley, New York.

COLEMAN, A. and MAGGS, K. R. A. (1965). *Land Use Survey Handbook*, An explanation of the Second Land Use Survey of Great Britain on the scale of 1:25,000, Isle of Thanet Geographical Association, Stanford, London.

COMMONWEALTH SCIENTIFIC AND INDUSTRIAL RESEARCH ORGANISATION (CSIRO), Australia (1967), *Lands of Bougainville and Buka Islands, Territory of Papua and New Guinea*. Land Research Series no. 20, Melbourne.

CONDON, R. W. (1968) 'Estimation of grazing capacity of arid grazing lands', in Stewart (1968) pp. 112–24.

COOKE, R. U., and HARRIS, D. R. (1970) 'Remote sensing of the terrestrial environment – principles and progress', *Transactions of the Institute of British Geographers*, **50**, 1–23.

CORNELL AERONAUTICAL LABORATORY INC. (1963) *Matrix methods for terrain profile simulation* (Project CATVAR) *see also* Deitchman.

COTTON, C. A. (1944) *Volcanoes as Landscape Forms*, Whitcomb & Tombs, New Zealand.

COWAN, W. L. (1956) 'Estimating hydraulic roughness coefficients', *Agricultural Engineering*, **37**, 473–5.

CSIRO. *See* Commonwealth Scientific and Industrial Research Organisation.

CURTIS, L. F., DOORNKAMP, J. C. and GREGORY, J. K. (1965) 'The description of relief in field studies of soils', *Journal of Soil Science*, **16**, no. 1, 16–30.

DALRYMPLE, J. R., BLONG, R. J. and CONACHER, A. J. (1968) 'A hypothetical nine unit landsurface model', *Zeitschrift für Geomorphologie*, **12**, 60–76.

DANSEREAU, P. (1958) *A Universal System for Recording Vegetation*. Institut Botanique de l'Université de Montreal, Canada.

DE CANDOLLE, A. P. (1856) *Géographie botanique raisonée*.

DEITCHMAN, S. J. (c [1966]) *Classification and quantitative description of large geographic areas to define transport systems requirements*, Cornell Aeronautical Laboratory.

DENNIS, J. (1835) *The Landscape Gardener*, James Ridgeway, London.

DICKINSON, R. E. (1930) 'The regional functions and zones of influence of Leeds and Bradford', *Geography*, **15**, 548–57.

DIRECTORATE OF OVERSEAS SURVEYS (1958) *Map of Gambia Land Use at 1 : 25,000 (approx.)*, DOS 3001.

DIRECTORATE OF SURVEY. WAR OFFICE AND AIR MINISTRY (1959) 1 : 250,000. Series K.5635. Edition 1-GSGS of SE Arabia.

DUFTON, A. F. (1940–41) 'Heat transmission coefficients', *Journal of the Institute of Heating and Ventilating Engineers*, **8**.

DUNCAN, O. D. (1961). *Statistical Geography*, Free Press, New York.

EDMONDS, D. T., PAINTER, R. B. and ASHLEY, G. D. (1970) 'A semiquantitative hydrological classification of soils in northeast England', *Journal of Soil Science*, **21**, no. 2, 256–64.

FAIRBRIDGE, R. W. (1946–47) 'Notes on the geomorphology of the Houtmans Abrolhos Islands', *Journal of the Royal Society of Western Australia*, **33**, 1–43.

FENNEMAN, N. M. (1916) 'Physiographic divisions of the United States', *Annals of the Association of American Geographers*, **6**, 19–98.

FENNEMAN, N. M. (1928) 'Physiographic divisions of the United States', *Annals of the Association of American Geographers*, **18**, 261–353.

GEIGER, R. (1965). *The Climate Near the Ground*, 3rd edn, Harvard University Press.

GEOLOGICAL SOCIETY OF AMERICA (annual) *Bibliography and Index of Geology exclusive of North America*; since 1969, this has also included

North America.

GEOMORPHOLOGICAL ABSTRACTS, ed. K. M. Clayton, published by the University of East Anglia.

GRABAU, W. E. and RUSHING, W. N. (1968) 'A computer-compatible system for quantitatively describing the physiognomy of vegetation assemblages', in Stewart (1968), pp. 263–75.

GRANT, K., ed. (1968) *Proceedings of Study Tour and Symposium on Terrain Evaluation for Engineering*, Division of Applied Geomechanics, CSIRO, Australia.

GRANT, K. (1970) *Terrain Classification for Engineering Purposes in the Marree area, South Australia*, CSIRO Division of Soil Mechanics Technical Paper no. 4.

GRANT, K. and AITCHISON, G. D. (1965) *An Engineering Assessment of the Tipperary area, Northern Territory, Australia*. Soil Mechanics Section, CSIRO, Australia.

GRANT, K. and LODWICK, G. D. (1968) *Storage and retrieval of information on a terrain classification system*. Paper no. 453T of the Symposium on Terrain Evaluation for Engineering, convened by the Division of Soil Mechanics, CSIRO, in association with the Fourth Conference of the Australian Road Research Board, Research Paper no. 117, Division of Soil Mechanics, CSIRO, Melbourne.

GRAY, D. M. (1965) 'Physiography characteristics and the runoff pattern', *Proceedings of the Hydrology Symposium* **4**, 321, National Research Council of Canada.

GREGORY, S. W. (1968). *Statistical Methods and the Geographer*, Longmans, London.

GRIGG, D. (1967) 'Regions, models, and classes', in Chorley and O. Haggett, eds, *Models in Geography*, Methuen, London, pp. 461–509.

GRIM, R. E. (1968). *Clay Mineralogy*, McGraw-Hill, New York.

HAGGETT, P. (1965) *Locational Analysis in Human Geography*, Edward Arnold, London.

HAGGETT, P. and CHORLEY, R. J. (1969) *Network Analysis in Geography*, Edward Arnold, London.

HAGOOD, M. J., DAMLEVSKY, N. D. and BEUM, C. O. (1941) 'An examination of the use of factor analysis in the problem of sub-regional delineation', *Rural Sociology*, **6**, 216–33.

HAMMOND, E. H. (1957) *Procedures in the Descriptive Analysis of Terrain*, Wisconsin.

HAMMOND, E. H. (1962) 'Land form geography and land form description', *The California Geographer*, **3**, 69–75.

HAMMOND, E. H. (1964) 'Classes of land-surface form in the forty eight states, U.S.A.', *Annals of the Association of American Geographers*, **54**, no. 1., Map supplement No. 4.

HARE, F. K. (1959) 'A photo-reconnaissance survey of Labrador-Ungava. Department of Mines and Technical Surveys, Canada', *Geographical Branch Memoir*, no. 6.

HEATH, G. R. (1956) 'A comparison of two basic theories of land classification and their adaptability to regional photo-interpretation key techniques', *Photogrammetric Engineering*, **22**, 144–68.

HERBERTSON, A. J. (1905) 'The major natural regions, an essay in systematic geography', *Geographical Journal*, **25**, 300–12.

HEYLIGERS, P. C. (1968) 'Quantification of vegetation structure on vertical aerial photographs', in Stewart (1968), pp. 251–62.

HILLS, G. A. (1942) 'An approach to land settlement problems in Northern Ontario', *Scientific Agriculture*, **23**, 212–16.

HILLS, G. A. (1949). *The Classification of N Ontario Lands According to their Potential for Agricultural Production*, Soil Report no. 1, Toronto, Ontario Department of Lands and Forests, Research Division.

HILLS, G. A. (1950) 'The use of aerial photographs in mapping soil sites', *Forestry Chronicle*, **4**, 37.

HILLS, G. A. and PORTELANCE, R. (1960) *The Glackmeyer Report on Multiple Land-use Planning*, Ontario Department of Lands and Forests.

HORTON, R. E. (1945) Erosional development of streams and their drainage basins: hydrophysical approach to quantitative morphology', *Bulletin of the Geological Society of America*, **56**, 275–370.

HOWARD, A. D. and SPOCK, L. E. (1940) 'Classification of landforms', *Journal of Geomorphology*, **3**, 332–45.

HOWARD, J. A. (1970a) 'Stereoscopic profiling of land-units from aerial photographs', *The Australian Geographer*, 259–68.

HOWARD, J. A. (1970b) *Multiband Concepts of forested Land-units*. Symposium on Photo-Interpretation. International Society of Photogrammetry, Commission 7, Dresden.

HUNTING TECHNICAL SERVICES LTD (1954 to present) *Reports on Soil Surveys* of Makhmour, Ishaqi, Nahrwan, Diyala and other areas in Iraq, Jebel Marra and Roseires in Sudan, Sind in West Pakistan, and other areas; *range classification* in Jordan; *land classification* in Ghana.

IGNATYEV, G. M. (1968) 'Classification of cultural and natural vegetation sites as a basis for land evaluation', Stewart (1968), pp. 104–111.

INSTITUTE OF HEATING AND VENTILATING ENGINEERS (1942) *The Computation of Heat Requirements for Buildings*.

ISARD, W. (1956) *Location and Space Economy: a general theory relating to industrial location, market areas, land use, trade, and urban structure*, Massachusetts Institute of Technology Press.

JACOBS, P. and WAY, D. (1969) *Visual Analysis of Landscape Development*, Graduate School of Design, Harvard University.

JENNY, H. (1958) 'Role of the plant factor in the pedogenic functions', *Ecology*, **39**, 5–16.

JOERG, W. L. G. (1914) 'Natural regions of northern America', *Annals of the Association of American Geographers*, **4**, 55–83.

JOHNSON, D. W. (1921) *Battlefields of the World War*, American Geographical Society Research Series no. 3, New York.

JOLY, F. (1957) 'Les milieus arides. Definition. Extension', *Notes marocaines*, **8**, 15–30.

KANTEY, B. A. and TEMPLER (1959) *Geotechnical map of the proposed trunk route Mariental-Asab, South West Africa*, Cape Town.

KANTEY, B. A. and WILLIAMS, A. A. B. (1962) 'The use of soil engineering maps for road projects', *Transactions of the South African Institute of Civil Engineers*, **4**, 149–59.

KAYE, C. A. (1957) 'Military geology in the US sector of the European theater of operations during World War II', *Bulletin of the Geological Society of America*, **68**, 47–54.

KIEFER, R. W. (1967) 'Terrain analysis for metropolitan fringe area planning', *Journal of the Urban Planning and Development Division, Proceedings of the American Society of Civil Engineers*, UP4, **93**, 119–39.

KING, L. (1962) *The Morphology of the Earth: a study and synthesis of world scenery*. Oliver & Boyd, Edinburgh.

KING, L. J. (1969) *Statistical Analysis in Geography*, Prentice-Hall, New York.

KLINGEBIEL, A. A. and MONTGOMERY, P. H. (1961). *Land capability classification*, US Department of Agriculture Soil Conservation Service, Agriculture Handbook no. 210.

KÖPPEN, W. (1931) *Grundriss der Klimakunde*, Walter de Gruyter, Berlin and Leipzig.

KRUMBEIN, W. C. and GRAYBILL, P. A. (1965) *An Introduction to Statistical Models in Geology*, McGraw-Hill, New York.

KUCHLER, A. W. (1949) 'A physiognomic classification of vegetation', *Annals of the Association of American Geographers*, **39**. 201–10.

LACATE, D. S., (1961) 'A review of land type classification and mapping', *Land Economics*, **37**, 271–8.

LANDSCAPE RESEARCH GROUP SYMPOSIUM (1967) *Methods of Landscape Analysis*, London: see Childs, D. R. Clouston, J. B., Mott, P. G., Tandy, C. R. V.

LANGBEIN, W. B. and HOYT, W. G. (1959) *Water Facts for the Nation's Future*, Ronald Press, New York.

LEWIS, P. H. '1964) 'Quality corridors for Wisconsin', *Landscape Architecture*, **54**, 2, 100–7.

LEOPOLD, L. B., WOLMAN, M. G. and MILLER, J. P. (1964) *Fluvial Processes in Geomorphology*. W. H. Freeman, San Francisco.

LINDLEY, D. V. and MILLER, J. C. P. (1968) *Cambridge Elementary Statistical Tables*, Cambridge University Press.

LINTON, D. L. (1951) 'The delimitation of morphological regions', in L. D. Stamp, and S. W. Wooldridge, eds., *London Essays in Geography*, Longmans, London, pp. 199–218.

LOBECK, A. K. (1923) *Physiographic Diagram of Europe*, Geography Press, Columbia University, New York.

LYON, T. L., BUCKMAN, H. O. and BRADY, N. C. (1952) *The Nature and Properties of Soils*, 5th edn, Macmillan, New York.

MABBUTT, J. A. (1968) 'Review of concepts of land classification', In Stewart (1968), pp. 11–28.

MABBUTT, J. A. and STEWART, G. A. (1965) 'The application of geomorphology in resources surveys in Australia and New Guinea', *Revue de Géomorphologie Dynamique*, July–Sept. nos. 7–8–9, pp. 97–109.

MACKENZIE, R. C., ed. (1957) *The Differential Thermal Investigation of Clays*, Mineralogical Society Clay Minerals Group, London.

MACKIN, J. H. (1948) 'Concept of the graded river', *Bulletin of the Geological Society of America*, **59**, 463–512.

MCHARG, I. (1967) 'Where should highways go? *Landscape Architecture*, **57**, 179–81.

MCNEIL, G. (1967) *Terrain Evaluation: data storage*. MEXE, Technical Note 5/67.

MEANS, R. E. and PARCHER, J. V. (1964) *Physical Properties of Soils*, Constable, London.

MEIER, R. L. (1965) *Development Planning*, McGraw-Hill, New York.

MEIGS, P. (1957) *World Distribution of Arid and Semi-arid Homoclimates*, with maps UN 392 and UN 393, Unesco, Paris.

MEINZER, O. E., ed. (1942) *Hydrology*, Dover Publications, New York.

MELTON, M. A. (1958) *List of sample parameters of quantitative properties of landforms: their use in determining the size of geomorphic experiments*, Technical Report no. 16, Office of Naval Research, Dept. of Geology, Columbia University.

MEXE. *See* Military Engineering Experimental Establishment,

MEYERHOFF, H. A. (1940) 'Migration of erosional surfaces'. *Annals of the Association of American Geographers*, **30**, 247–54.

MILITARY ENGINEERING EXPERIMENTAL ESTABLISHMENT (1965) *The classification of terrain intelligence*. Reports of the Combined Pool (AER) 1960–64. MEXE Report no. 915.

MILITARY ENGINEERING EXPERIMENTAL ESTABLISHMENT (1968) *User Handbook for the Soil Assessment Cone Penetrometer*, Army Code no. 60285 (provisional handbook (1966) was Army Code no. 14554).

MILL, J. S. (1891) *A System of Logic*, 8th edn, Harper New York, as quoted in Soil Survey Staff, US Department of Agriculture (1960), *Soil classification. A comprehensive system. 7th Approximation*.

MILLER, A. A. (1946) *Climatology* Methuen, London.

MILLER, A. A. (1965) *The Skin of the Earth*, Methuen University Paperback, London.

MILLER, R. L. and KAHN, R. S. (1962) *Statistical Analysis in the Geological Sciences*. J. Wiley, New York.

MILLER, T. G. (1967) 'Recent studies in military geography', (review article) *Geographical Journal*, **133**, 354–6.

MILNE, G. (1935) 'Some suggested units of classification and mapping, particularly for East African soils', *Soil Research*, **4**, no. 3, 183–98.

MINISTRY OF AGRICULTURE, FISHERIES, AND FOOD (1968) *Agricultural*

Classification map of England and Wales.

MINISTRY OF OVERSEAS DEVELOPMENT (UK), Land Resources Division, Directorate of Overseas Surveys. *Assistance to Developing Countries. Planning the use of land resources.*

MITCHELL, C. W. and PERRIN, R. M. S. (1966) 'The subdivision of hot deserts of the world into physiographic units', *Actes du IIe Symposium International de Photo-Interprétation,* Sorbonne, Paris, IV, 1–89 to IV, 1–106.

MORE, R. J. (1967) 'Hydrological models in geography'. In R. J. Chorley, and P. Haggett, eds, *Models in Geography.* Methuen, London.

MORONEY, M. J. (1956). *Facts from Figures.* 3rd edn, Penguin Books (Pelican) Harmondsworth.

MORSE, P. K. and THORNBURN, T. H. (1961) 'Reliability of soil maps', *Proceedings of the 5th International Conference of Soil Mechanics and Foundation Engineering,* pp. 259–62.

MOTT, P. G. (1967) 'Air photography as an aid to experimental planning', in Landscape Research Group (1967), pp. 31–6.

MOTTS, W. S., ed. (1970) *Geology and Hydrology of Related Playas in Western United States.* University of Massachusetts, Amherst, Final scientific report for Air Force Cambridge Research Laboratories.

MOUNTAIN, M. J. (1964) *Soils engineering map of an area in the immediate vicinity of the Etosha Pan, Ovamboland, South West Africa,* Kantey and Templer, Cape Town.

MURPHY, R. E. (1968) 'Landforms of the world', *Annals of the Association of American Geographers,* **58**, no. 1; Map Supplement no. 9.

MURRAY, A. C. (1967) *Power Station Siting: visual analysis.* in Landscape Research Group, *Methods of landscape analysis,* pp. 20–3.

NAKANO, T. (1962) 'Landform type analysis on aerial photographs, its principle and techniques. *Archives Internationales de Photogrammetrie,* **14**, *Transactions of the Symposium on Photo-Interpretation,* Delft, pp. 149–52.

OFFICE OF NAVAL RESEARCH (US NAVY): *see* Melton, M. A.

OLIVIER, H. (1961) *Irrigation and Climate,* Edward Arnold, London.

OLSON, C. E. (1970). *Multi-spectral Remote Sensing,* a report to Commission VII of the International Society of Photogrammetry at the Dresden Symposium on Photo-Interpretation.

OLSON, C. E. (1971) 'Collection and processing of multi-spectral imagery', in International Union of Forest Research Organisations, Section 25, Joint Report by Working Group. *Application of Remote Sensing in Forestry,* pp. 7–20.

PARRY, J. T., HEGINBOTTOM, J. A. and COWAN, W. R. (1968) Terrain analysis in mobility studies for military vehicles', in Stewart (1968), Australia, 160–70.

PARRY, M. (1971) Private communication.

PASSARGE, S. (1919) *Die Grundlagen der Landschaftskunde.* Hamburg.

PENCK, W. (1927) *Die morphologische Analyse,* Stuttgart.

PENMAN, H. L. (1963) *Vegetation and Hydrology*. Commonwealth Bureau of Soils, Harpenden, Technical Communication no. 53.

PERRIN, R. M. S. and MITCHELL, C. W. (1970) *An appraisal of physiographic units for predicting site conditions in arid areas*, MEXE Report no. 1111, 2 vols.

PHILLIPS, E. (1965) *Field Ecology; a laboratory block*, American Institute of Biological Sciences, Heath, Boston.

PHOTOGRAPHIC SURVEY CORPORATION LTD, Toronto, Canada (1956). *Landforms and soils, West Pakistan*, scale 1:253 440. Produced in cooperation with Central Soil Conservation Organization, Ministry of Food and Agriculture, Pakistan.

PHOTO INTERPRÉTATION (1961) *Editions Technip*. 7, Rue Nelaton, Paris, 15e.

POORE, M. D. (1956) 'The use of phytosociological methods in ecological investigations. iv. General discussion of phytosociological problems', *Journal of Ecology*, **44**, 28–50.

PROKAIEV, V. I. (1962) 'The facies as a basic and smallest unit in landscape studies', *Soviet Geography, Review and Translation*, **3**, no. 6, 21–9.

PUTNAM, W. C., AXELROD, D. I., BAILEY, H. P., and MCGILL, J. T. (1960), *Natural Coastal Environments of the World*. University of California, Los Angeles.

QUARTERMASTER RESEARCH AND ENGINEERING CENTRE (US ARMY): *see* Anstey, F.; Wood, W. F. and Snell, J. B.

RAISZ, E. J. (1938) 'Developments in the physiographic method of representing the landscape on maps', *Proceedings of the 15th International Geographical Congress, Amsterdam*, **2**, section 1, 140–9.

RAISZ, E. J. (1946) 'Landform, landscape, land use, and land type maps', *Annals of the Association of American Geographers*, **36**, no. 1, 102–3.

RENWICK, C. C. (1968) 'Land assessment for regional planning: the Hunter region of N.S.W. as a case study in land evaluation'. in Stewart (1968), pp. 171–9.

REPUBLIC OF THE SUDAN MINISTRY OF AGRICULTURE (1963) *Roseires Soil Survey*. Report No. 1. *Gezira Extension Area. Soil Survey and Land Classification*, vol. 1, Sir Murdoch Macdonald and Partners by Hunting Technical Services Ltd.

ROBERTSON, V. C., JEWITT, T. N., FORBES, A. P. S. and LAW, R. (1968) 'The assessment of land quality for primary production', in Stewart (1968), pp. 88–103.

ROBERTSON, V. C. and STONER, R. F. (1970) 'Land use surveying: a case for reducing the costs', in *World Land Use Survey Occasional Papers* no. 9, Geographical Publications, Berkhamsted, pp. 3–15.

ROBINSON, G. W. (1949) *Soils, Their Origin, Constitution, and Classification*, Thomas Murby, London.

ROSENFELD, A. (1968) 'Automated picture interpretation', in Stewart (1968), pp. 187–99.

ROXBY, P. M. (1926) 'The theory of natural regions', *The Geographical Teacher*, **13**, 376–82.

SAINT ONGE, D. A. (1968) 'Geomorphic maps', in R. W. Fairbridge, *Encyclopedia of Geomorphology*, Reinhold, New York.

SAMPFORD, M. R. (1962) *An Introduction to Sampling Theory*, Oliver & Boyd, Edinburgh.

SAVIGEAR, R. A. G. (1952) 'Some observations on slope development in South Wales', *Transactions and Papers of the Institute of British Geographers*, **18**, 31–51.

SAVIGEAR, R. A. G. (1956) 'Techniques and terminology in the investigation of slope forms', *Premier Rapport de la Commission pour l'Etude des Versants, Union Géographique Internationale*, pp. 66–75.

SAVIGEAR, R. A. G. (1960) 'Slopes and hills in West Africa', *Zeitschrift für Geomorphologie*, N.F., Suppl. vol. 1, pp. 156–71.

SAVIGEAR, R. A. G. (1962) 'Some observations on slope development in North Devon and North Cornwall', *Transactions and Papers of the Institute of British Geographers*, **31**, 23–42.

SAVIGEAR, R. A. G. (1965) 'A technique of morphological mapping', *Annals of the Association of American Geographers*, **55**, no. 3, 514–38.

SCHIMPER, A. F. W. (1903) *Plant Geography upon a Physiological Basis*, Oxford.

SCHNEIDER, S. J. (1966) 'The contribution of geographical air-photo interpretation to problems of land division according to natural units', *Actes du IIe Symposium International de Photo-Interpretation*, Paris, VI–23 to VI–28.

SCHWAB, G. O., FREVERT, R. K., EDMINSTER, T. W. and BARNES, K. B. (1966) *Soil and Water Conservation Engineering*, J. Wiley, New York.

SCOTT, R. M., HEYLIGERS, P. B., MCALPINE, J. R., SAUNDERS, J. C. and SPEIGHT, J. G. (1967) *Lands of Bougainville and Buka Islands, Territory of Papua and New Guinea*, CSIRO Land Research Series no. 20.

SHERMAN, L. K. (1932) Streamflow from rainfall by unit hydrograph method', *Engineering News-Record*, **108**, 501–5.

SIEGAL, S. (1956) *Non-parametric Statistics for the Behavioural Sciences*, McGraw-Hill, New York.

SMITH, J. (1949) *Distribution of tree species in the Sudan in relation to rainfall and soil texture*, Sudan Government Ministry of Agriculture, Bulletin no. 4.

SMITH, K. G. (1950) 'Standards for grading texture of erosional topography' *American Journal of Science*, **648**, 655–68.

SNEDECOR, G. W. and COCHRAN, W. G. (1967) *Statistical Methods*, 6th edn, Iowa State University Press, Ames.

SOIL SURVEY STAFF (1951) *Soil Survey Manual*, US Department of Agriculture Handbook no. 18.

SOIL SURVEY STAFF (1960) *Soil classification: a comprehensive system. 7th Approximation*, US Department of Agriculture.

SOLNTSEV, N. A. (1962) 'Basic problems of Soviet landscape science',

Soviet Geography, Review and Translation, **3**, no. 6, 3–15.

SOUTH AFRICAN INSTITUTE OF CIVIL ENGINEERS (1964 and 1965). *Proceedings of the First and Second Symposia on Soil Engineering Mapping and Data Storage*, La Hermosa, Transvaal, and Magaliesburg, Transvaal respectively, ed. A. B. A. Brink and T. C. Partridge.

SPRUNT, B. F. (1970) 'Geographics: a computer's eye view of terrain', *Area (Institute of British Geographers)*, **4**, 54–9.

STAMP, L. D. and WILLATTS, E. C. (1935) *The Land Utilization Survey of Britain*, London School of Economics.

STEIN, A. (1959) 'Terrain mark angles', *Joint session of the AAS Section E and Geological Society of America Symposium on quantitative terrain studies*, Chicago.

STEINITZ, C. F. (1970). 'Landscape resource analysis: the state of the art', *Landscape Architecture*, **60**, **2**, 101–4.

STEWART, G. A., ed. (1968) *Land Evaluation*, Macmillan of Australia.

STONE, R. O. and DUGUNDJI, J. (1965) 'A study of micro-relief, its mapping, classification, and quantification by means of Fourier analysis', *Engineering Geology*, **1**, no. 2, 89–187.

STRAHLER, A. N. (1957) 'Quantitative analysis of watershed geomorphology', *Transactions of the American Geophysical Union*, **38**, 913–20.

STRAHLER, A. N. (1964), in Chow, Ven te, ed. (1964).

STRAHLER, A. N. (1967) *Introduction to Physical Geography*, Wiley, New York.

TA LIANG. (1964) *Tropical Soils: characteristics and airphoto interpretation*, School of Civil Engineering, Cornell University; prepared for Air Force Cambridge Research Laboratories.

TANDY, C. R. V. (1967) 'The isovist method of landscape survey', in Landscape Research Group (1967), pp. 9–10.

TANSLEY, A. G. (1953) *The British Islands and Their Vegetation*, 2 vols, Cambridge University Press.

TAYLOR, B. W. (1959) 'Ecological land use surveys in Nicaragua', *Estudios Ecologicos*, vol. 1.

THIRLAWAY, H. I. S. (1959) *Annual Report of the Unesco Arid Zone Geophysical Research Project in Pakistan*.

THORNBURN, T. H. and LARSEN, W. R. (1959) 'A statistical study of soil sampling', *Journal of the Soil Mechanics and Foundation Engineering Division of the Proceedings of the American Society of Civil Engineers*, **85**, SM5, pp. 1–13.

THORNBURY, W. D. (1954) *Principles of Geomorphology*, Wiley, New York.

THORNTHWAITE, C. W. (1933) 'The climates of the earth', *Geographical Review*, **21**, 633–55.

THORNTHWAITE, C. W. (1948) 'An, approach towards a rational classification of climate, *Geographical Review*, **38**, 85–94.

THORNTHWAITE, C. W. (1954) 'A re-examination of the concept and measurement of practical transpiration, in J. R. Mather, ed., *The Measurement of Potential Evapotranspiration: problems in climatology*, Seabrook, New Jersey, 200–9.

TOMLINSON, R. F. (1968) 'A geographic information system for regional planning; in Stewart (1965), pp. 200–10.

TOMLINSON, R. F. (1971) Private communication.

TOTHILL, J. D., ed. (1952) *Agriculture in the Sudan*, Oxford University Press.

UNITED NATIONS (1971) Food and Agriculture Organisation (FAO) *Soil map of the world.*

USAEWES: *see* United States Army Engineer Waterways Experiment Station.

UNITED STATES ARMY ENGINEER WATERWAYS EXPERIMENT STATION (1959) *Handbook. A technique for preparing desert terrain analogs*, Technical Report no. 3–506.

UNITED STATES ARMY ENGINEER WATERWAYS EXPERIMENT STATION (1961) *Trafficability of soils. Soil classification*, Technical Memorandum no. 3–240.

UNITED STATES ARMY ENGINEER WATERWAYS EXPERIMENT STATION (1962) 'Desert terrain evaluations. $\frac{1}{4}$ ton utility truck at Yuma' (draft).

UNITED STATES ARMY ENGINEER WATERWAYS EXPERIMENT STATION (1963a) *Military evaluation of geographic areas. Reports on activities to 1963*, Miscellaneous Paper no. 3–610.

UNITED STATES ARMY ENGINEER WATERWAYS EXPERIMENT STATION (1963b) *Forecasting trafficability of soils*, Technical Memorandum no. 3–331, vol. 2.

UNITED STATES ARMY ENGINEER WATERWAYS EXPERIMENT STATION (1963c) *Environmental factors affecting ground mobility in Thailand*, Technical Report no. 5–625, Appendix C, Trafficability.

UNITED STATES ARMY ENGINEER WATERWAYS EXPERIMENT STATION (1965–67) *Terrain analysis by electromagnetic means*, Technical Report no. 3–693, Reports 1–4.

UNITED STATES ARMY ENGINEER WATERWAYS EXPERIMENT STATION (1968) *Mobility environmental research study, quantitative method for describing terrain for ground mobility*. Technical Report no. 3–726.

UNITED STATES DEPARTMENT OF AGRICULTURE (1954) *Diagnosis and Improvement of Saline and Alkali Soils*, Agriculture Handbook no. 60.

UNITED STATES DEPARTMENT OF THE INTERIOR, Bureau of Reclamation (1951) *Manual*, vol. 5, *Irrigated Land Use*, Denver, Colorado.

UNITED STATES GEOLOGICAL SURVEY (annual) *Bibliography of North American Geology.*

UNSTEAD, J. F. (1933) 'A system of regional geography', *Geography*, **18**, 175–87.

UNSTEAD, J. F. *et al.* (1937) 'Classification of the regions of the world, *Geography*, **22**, 253–82.

VAN BEERS, W. F. J. (1958) *The Auger Hole Method for Field Measurement of Hydraulic Conductivity*, International Institute for Land Reclamation and Improvement, Wageningen, Bulletin no. 1.

VAN ROESSEL, J. (1971) 'Automated mapping of forest resources from digitised aerial photographs', in International Union of Forest

Research Organisations Section 25, *Joint Report by Working Group. Application of remote sensors in foresty*, pp. 177–88.

VAN VALKENBURG, S. A. and HUNTINGTON, E. (1935) *Europe*, Wiley, New York.

VEATCH, J. O. (1933) *Agricultural Classification and Land Types of Michigan*, Michigan Agricultural Experiment Station, Special Bulletin no. 231.

VERSTAPPEN, H. TH. and VAN ZUIDAM, R. A. (1968) 'ITC system of geomorphological survey', in *ITC Textbook of photo-interpretation*, Delft, chapter 7.

VINOGRADOV, B. V., GERENCHUK, K. I., ISACHENKO, A. G., RAMAN, K. G. and TSESELCHUK, YU N. (1962), 'Basic principles of landscape mapping', *Soviet Geography, Review and Translation*, **3**, no. 6, 15–20.

VON ENGELN, D. D. (1942) *Geomorphology*, Macmillan, New York.

WALLACE, A. R. (1876) *The Geographical Distribution of Animals*, Macmillan, London.

WATERS, R. S. (1958) 'Morphological mapping', *Geography*, **43**, 10–17.

WAY, D. (1968) *Airphoto Interpretation for Land Planning*, Harvard University Department of Landscape Architecture.

WEBSTER, R. and BECKETT, P. H. T. (1964) 'A study of the agronomic value of soil maps interpreted from air photographs', *Transactions of the 8th International Congress of Soil Science*, **5**, 795–803.

WEBSTER, R. and BECKETT, P. H. T. (1968) 'Quality and usefulness of soil maps', *Nature, Lond*, **219**, 680–2.

WEBSTER, R. and BECKETT, P. H. T. (1970) 'Terrain classification and evaluation using air photography: a review of recent work at Oxford', *Photogrammetria*, **26**, 51–75.

WHITTLESEY, D. (1936) 'Major agricultural regions of the earth', *Annals of the Association of American Geographers*, **26**, 199–240.

WOOD, W. F. and SNELL, J. B. (1957) *The Dispersion of Geomorphic Data Around Measures of a Central Tendency and its application*, HQ Quartermaster Research and Engineering Command, Technical Report EA–8, Natick, Mass.

WOOD, W. F. and SNELL, J. B. (1959) *Predictive Methods in Topographic analysis. I. Relief, slope, and dissection on inch-to-the-mile maps in the USA*, HQ Quartermaster Research and Engineering Command, Technical Report EP–112, Natick, Mass.

WOOD, W. F. and SNELL, J. B. (1960) *A Quantitative System for Classifying Landforms*, HQ Quartermaster Research and Engineering Command, Technical Report EP–124, Natick, Mass.

WOOLDRIDGE, S. W. (1932) 'The cycle of erosion and the representation of relief', *Scottish Geographical Magazine*, **48**, 30–6.

WRIGHT, R. L. (1964), *Unesco Geomorphology Mission, Pakistan*, Report to Unesco.

WRIGHT, R. L. (1967) 'A geomorphological approach to land classification, Ph.D thesis', University of Sheffield.

YATES, F. (1960) *Sampling Methods for Censuses and Surveys*, 3rd ed, Griffin, London.

YOUNG, A. (1963) 'Some field observations on slope form and regolith and their relation to slope development', *Transactions and Papers of the Institute of British Geographers*, **32**, 1–29.

YOUNG, A. (1971) 'Slope profile analysis: the system of best units', in Institute of British Geographers, *Slopes: form and process*, Special Publication no. 3, pp. 1–13.

YULE, G. G. and KENDALL, M. G. (1950) *An Introduction to the Theory of Statistics*, 14th ed, Griffin, London.

Index